PUBLISHERS' NOTE

The series of monographs in which this title appears was introduced by the publishers in 1957, under the General Editorship of Dr Maurice G. Kendall. Since that date, more than twenty volumes have been issued, and in 1966 the Editorship passed to Alan Stuart, D.Sc.(Econ.), Professor of Statistics, University of London.

The Series fills the need for a form of publication at moderate cost which will make accessible to a group of readers specialized studies in statistics or courses on particular statistical topics. Often, a monograph on some newly developed field would be very useful, but the subject has not reached the stage where a comprehensive treatment is possible. Considerable attention has been given to the problem of producing these books speedily and economically.

It is intended that in future the Series will include works on applications of statistics in special fields of interest, as well as theoretical studies. The publishers will be interested in approaches from any authors who have work of importance suitable for the Series.

CHARLES GRIFFIN & CO. LTD.

GRIFFIN'S STATISTICAL MONOGRAPHS AND COURSES

GENERALIZED INVERSE MATRICES WITH APPLICATIONS TO STATISTICS

R. M. PRINGLE

B.Sc. Agric. (Natal), Ph.D. (Natal)
Senior Lecturer in Biometry, University of Natal
and Department of Agricultural Technical Services,
South Africa

and

A. A. RAYNER

M.A. (New Zealand), Ph.D. (Edinburgh)
Professor of Biometry, University of Natal
and Department of Agricultural Technical Services,
South Africa

BEING NUMBER TWENTY-EIGHT OF
GRIFFIN'S STATISTICAL
MONOGRAPHS & COURSES
EDITED BY
ALAN STUART, D.Sc.(Econ.)

GRIFFIN LONDON

CHARLES GRIFFIN & COMPANY LIMITED
42 DRURY LANE, LONDON, WC2B 5RX

First published 1971
ISBN: 0 85264 181 8

Set by E W C Wilkins & Associates Ltd London N12 OEH
Printed in Great Britain by Latimer Trend & Co Ltd Whitstable.

PREFACE

This monograph is based on a thesis by the first author, approved for the degree of Doctor of Philosophy in the University of Natal in February, 1969.

The second author, who acted as supervisor for the thesis, first became interested in the subject of the "inverse" of a singular matrix from his reading of the classical paper of Yates and Hale (1939). In 1955, when extending results on the distribution and independence of quadratic forms in normal variates to the case when the variance matrix of the variates is singular (cf. Rayner and Livingstone, 1965), he realized that this concept must be highly relevant, and he can remember studying the paper of Penrose (1955) on its first appearance without, however, deriving any benefit from it at the time. Later, while on sabbatical leave at the University of North Carolina in 1962–63, he found that Professors R.C. Bose and H.L. Lucas were using in their lecture notes what they called a "conditional inverse", and he was privileged to have a number of discussions with C.A. Rohde, who later produced a thesis on generalized inverses. At about the same time Rao (1962) linked a generalized inverse of a matrix with the distribution of a quadratic form in singular normal variates.

Such has been the increasing interest among both statisticians and mathematicians in the subject of generalized inverses that a symposium devoted to this single topic was held in Texas in March, 1968. This interest among both statisticians and mathematicians is partly responsible for the diversity of mathematical and statistical journals in which papers on generalized inverses have appeared, for considerable duplication of published results, for varying definitions, and for a confusing range of notation and nomenclature. In these circumstances, we considered that a cohesive account in monograph form of the mathematical properties and statistical applications of generalized inverses might well have merit.

In accordance with this broad objective, we have divided the book into two sections, mathematical and statistical, and in doing so we have frankly motivated the mathematical part towards the statistical applications, rather than present an exhaustive treatment. For example, we consider only matrices in the real field, although several of the papers we refer to do not have this restriction; and we do not go into much detail in respect of the purely numerical aspects of the calculation of generalized inverses. We assume that the reader will have a reasonably thorough knowledge of elementary matrix algebra such as that contained in the books of Aitken (1956) or Searle (1966).

In view of the topical nature of the subject, it is desirable to draw attention to the fact that some of the new results of the first author's thesis incorporated in this monograph have been independently obtained at approximately the same time by the Indian school of C.R. Rao, S.K. Mitra, and C.G. Khatri. We also record that, apart from an isolated paper (Chipman, 1968), we have not yet seen the proceedings of the Texas symposium. We acknowledge fully our debt to the thesis of C.A. Rohde (1964), which was the starting-point of this work.

In conclusion, it is a pleasure to acknowledge our indebtedness to Professor J.S. Chipman of the University of Minnesota for an interesting and helpful correspondence on a number of points, and to Drs B.M. Nevin and N.C.K. Phillips of the Mathematics Department, and Miss I.M. Gravett of the Biometry Department, of the University of Natal for helpful discussions. We also thank Professor G. Zyskind of Iowa State University for making available a copy of Zyskind & Martin (1967) prior to publication, and Professor R.L. Plackett of the University of Newcastle-upon-Tyne for his suggestion as regards publication. Our final acknowledgement is to Mrs S.E. Hancock for her conscientious and masterly typing of the manuscript.

R.M. PRINGLE

Pietermaritzburg, A.A. RAYNER

November, 1969

CONTENTS

Chapter *Page*

Chapter 1

DEFINITIONS AND SOME PRELIMINARY RESULTS

1.1 Notation and general introduction

In this monograph we consider only real matrices, which are denoted by ordinary italic capitals such as A, B, etc. Italic lower-case letters will denote scalars, while bold-face lower-case letters will be used to represent vectors.

The concept of a generalized inverse matrix has its roots in the theory of simultaneous linear equations. The solution of a set of consistent linear equations

$$A\mathbf{x} = \mathbf{h}, \tag{1.1}$$

where A is $n \times k$ of rank $r \leqslant \min(n, k)$, may assume two different forms. If $n = k = r$, a unique solution $\mathbf{x} = A^{-1}\mathbf{h}$ exists. However, when A is rectangular or square singular, a simple representation of a solution in terms of A is more difficult, and recourse has been made by Penrose (1955), Bjerhammar (1958), Bose (1959), and others to the use of generalized inverse matrices. Broadly speaking, a generalized inverse matrix of A is some matrix A^* such that $A^*\mathbf{h}$ is a solution to [1.1].

Although the first published work on generalized inverses dates back to Moore (1920), little investigation of the theoretical properties of such matrices was undertaken until 1955, when Penrose defined a uniquely determined generalized inverse matrix and investigated some of its properties. Penrose defined his generalized inverse as follows:

For any matrix A, square or rectangular, there exists a *unique* matrix G satisfying the conditions

$$\left.\begin{array}{ll} (1) & AGA = A \\ (2) & GAG = G \\ (3) & (AG)' = AG \\ (4) & (GA)' = GA. \end{array}\right\} \tag{1.2}$$

However, as Penrose himself pointed out, a solution to [1.1] requires a generalized inverse which satisfies only condition (1), and it was on this type of generalized inverse that most of the research of the late 1950's and early 1960's was concentrated. Workers such as Rao (1955, 1962), Bjerhammar (1958), Wilkinson (1958), and Bose (1959), dispensing with the uniqueness requirement, exploited the flexibility of this weaker type of generalized inverse in statistical applications, especially the analysis of the linear model. For their purposes the lack of uniqueness was no loss, since they found that certain results in least squares theory, in particular the estimates of estimable linear functions and the variances of these estimates, were invariant under the choice of generalized inverse.

There are two further types of generalized inverses useful in statistics. The first of these, a matrix which satisfies conditions (1) and (2) of [1.2], was considered by Rao (1955, 1962) and also by Bjerhammar (1958). The second variant, studied by Goldman & Zelen (1964), was required to satisfy conditions (1), (2), and (4) of [1.2]. For notational reasons associated with statistical applications it will more often be convenient to use a generalized inverse satisfying conditions (1), (2), and (3) rather than (1), (2), and (4). Algebraically, however, there is little difference, and results involving this class of generalized inverse will often be given in either or both forms.

It may be mentioned that the historical development of the subject is not always clear. This is largely because the work of R.C. Bose, although available in mimeographed form, has never been published, and also because many of Bjerhammar's papers are either inaccessible or have been published in Swedish.

The following system of notation and nomenclature for the four types of generalized inverse matrices briefly introduced above will be used in this monograph. It is assumed for the purposes of this table and for all future references that conditions (1), (2), (3), and (4) relate to those expressed in [1.2].

Conditions satisfied	Name	Abbreviation	Notation
(1)	One-condition generalized inverse	g_1-inverse	A^{g_1}
(1) & (2)	Two-condition generalized inverse	g_2-inverse	A^{g_2}
(1), (2), & (3)	Three-condition generalized inverse	g_3-inverse	A^{g_3}
(1), (2), & (4)		g_3^*-inverse	$A^{g_3^*}$
(1), (2), (3) & (4)	The generalized inverse	The g-inverse	A^g

Note that "*the* g-inverse" of A refers to A^g, whereas "*a* g-inverse" of A may be used for brevity to refer to any generalized inverse of A.

The terms proposed in the above table are used inclusively; for example, a particular g_1-inverse may actually be a g_2-inverse, but there may be no need to utilize the second condition. This aspect will be discussed further at a later stage.

It is interesting to note that Chipman (1968) has proposed a similar and even more explicit system of notation for the various classes of generalized inverses.

1.2 The generalized inverse

Moore (1935) investigated the concept of a "general reciprocal" of a matrix, first introduced by him in 1920. According to Greville (1962) (Moore used a very personal and complex system of notation), Moore's definition of the general reciprocal of a matrix A is the matrix G which satisfies condition (1) and also has its row-space and column-space contained in the row-space and column-space, respectively, of A'. In 1955 Penrose, independently and through a different approach, discovered Moore's general reciprocal, which he called the "generalized inverse". Since Penrose's approach is more in keeping with the familiar matrix techniques which will be employed, this type of g-inverse will be introduced in his manner.

DEFINITION 1.1 The g-inverse of an $n \times k$ matrix A is defined as the $k \times n$ matrix $G = A^g$ for which equations [1.2] have a unique solution.

This definition clearly includes the case $A = 0$, for the g-inverse of a null matrix is equal to its transpose. The relationship between the work of Moore and Penrose has been discussed by Rado (1956) and Greville (1962).

Greville (1959) and Rohde (1964, p. 16) prefer to use the name "pseudo-inverse" when referring to A^g, presumably because such a matrix behaves almost like an ordinary inverse. Another name which is frequently used (e.g. Rao, 1965, p. 25) is the "Moore–Penrose inverse".

The existence of A^g will be established by the following orthonormal reduction of A to diagonal form:

LEMMA Let A be an $n \times k$ matrix of rank r, and let λ_i be the ith nonzero latent root of AA'; then there exist row-orthonormal matrices H_1 and Q (i.e. $H_1 H_1' = I$, $QQ' = I$) such that

$$A = H_1' \Lambda^{\frac{1}{2}} Q, \qquad [1.3]$$

where $\Lambda^{\frac{1}{2}} = \text{diag}(\lambda_1^{\frac{1}{2}}, \lambda_2^{\frac{1}{2}}, \ldots, \lambda_r^{\frac{1}{2}})$.

Proof: Since AA' is real symmetric, there exists H orthogonal such that

$$HAA'H' = \begin{bmatrix} \Lambda & 0 \\ 0 & 0 \end{bmatrix}.$$

Let H be partitioned as

$$\begin{bmatrix} H_1 \\ H_2 \end{bmatrix},$$

where H_2 corresponds to the $n - r$ zero latent roots of AA'; then

$$H_1 AA'H_1' = \Lambda, \qquad [1.4]$$

$$H_2 A = 0, \qquad [1.5]$$

and

$$H_1'H_1 = I - H_2'H_2. \qquad [1.6]$$

If $\Lambda^{-\frac{1}{2}}H_1 A = Q$, then $QQ' = \Lambda^{-\frac{1}{2}}H_1 AA'H_1'\Lambda^{-\frac{1}{2}} = I_r$ by [1.4]. Also, by [1.5] and [1.6],

$$\begin{aligned} A = (I - H_2'H_2)A &= H_1'H_1 A \\ &= H_1'\Lambda^{\frac{1}{2}}Q. \end{aligned}$$

The proof of the lemma has been adapted from Hsu (1946). Alternative proofs have been presented by Lanczos (1958) and Chipman & Rao (1964), who also established the existence of A^g in the following way:

THEOREM 1.1 Equations [1.2] have a unique solution for any A.

Proof: By the lemma it is always possible to express the matrix A in the form [1.3]. Consider the matrix $G = Q'\Lambda^{-\frac{1}{2}}H_1$. Then

$$\begin{aligned} AG &= H_1'\Lambda^{\frac{1}{2}}QQ'\Lambda^{-\frac{1}{2}}H_1 = H_1'H_1 \quad \text{(symmetric)}, \\ GA &= Q'\Lambda^{-\frac{1}{2}}H_1H_1'\Lambda^{\frac{1}{2}}Q = Q'Q \quad \text{(symmetric)}, \\ AGA &= H_1'H_1H_1'\Lambda^{\frac{1}{2}}Q = A \end{aligned}$$

and

$$GAG = Q'\Lambda^{-\frac{1}{2}}H_1H_1'H_1 = G.$$

Thus G is a solution. Following Penrose, uniqueness is shown by noting that if U also satisfies [1.2], then:

$$\begin{aligned} A^g &= A^g\underline{AA^g} = A^g(A^g)'A'\underline{U'A'} = A^g\underline{(A^g)'A'}AU \\ &= A^gA\underline{U} = \underline{A^gA}A'U'U = \underline{A'U'}U = UAU = U, \end{aligned}$$

where the relations [1.2] have been freely used.

The reduction [1.3] was used by Rao (1966) to present a unique spectral decomposition of a rectangular matrix A. The spectral representation in turn leads to an alternative representation for A^g. Let $\mathbf{e}_i$ denote the ith unity column vector (Bodewig, 1956, p. 2); then [1.3] may be written as

$$A = \sum_{i=1}^{r} \lambda_i^{\frac{1}{2}} H_1' \mathbf{e}_i \mathbf{e}_i' Q.$$

Writing V_i for $H_1' \mathbf{e}_i \mathbf{e}_i' Q$, we have

$$A = \sum_{i=1}^{r} \lambda_i^{\frac{1}{2}} V_i$$

and

$$A^g = G = \sum_{i=1}^{r} \lambda_i^{-\frac{1}{2}} V_i'. \qquad [1.7]$$

Similar results were obtained by Penrose and by Chipman & Rao. It is easily verified from the definition of V_i that $V_i' = V_i^g$, and thus

$$A^g = \sum_{i=1}^{r} \lambda_i^{-\frac{1}{2}} V_i^g.$$

An alternative proof of the existence of A^g has been given by Greville (1960), who mentions that his result is based upon an idea suggested by A.S. Householder. Greville uses the result that any $n \times k$ matrix A of rank r may be factored as $A = BC$, where B is $n \times r$, C is $r \times k$, and both B and C have rank r. (An example is to let $B = H_1' \Lambda^{\frac{1}{2}}$ and $C = Q$ in [1.3](*).) Clearly $B^g = (B'B)^{-1}B'$ and $B^g B = I$. Similarly $C^g = C'(CC')^{-1}$ and $CC^g = I$. Using this factorization of A, Greville expresses A^g as

$$A^g = C'(CC')^{-1}(B'B)^{-1}B' = C^g B^g. \qquad [1.8]$$

The conditions under which a "reverse order law" holds for the g-inverse of a matrix product will be discussed in § 2.7. A further proof of the existence of A^g has been presented by Albert & Sittler (1966).

Penrose studied several properties of the g-inverse. Some of his results are summarized in the following theorem; the remainder will be discussed under the appropriate headings in Chapter 2.

(*) For further results on this factorization see Searle (1966, p. 119).

THEOREM 1.2

(a) $(A^g)^g = A$.

(b) $(A')^g = (A^g)'$.

(c) $A^g = A^{-1}$ if A is nonsingular.

(d) $A = AA'(A^g)' = (A^g)'A'A$.

(e) $A^g = A^g(A^g)'A' = A'(A^g)'A^g$.

(f) $(A'A)^g = A^g(A^g)'$.

(g) If $A = \Sigma A_i$, where $A_i' A_j = 0$ and $A_i A_j' = 0$ $(i \neq j)$, then $A^g = \Sigma A_i^g$.

(h) $(\alpha A)^g = \alpha^g A^g$, where α is any scalar with $\alpha^g = \alpha^{-1}$ if $\alpha \neq 0$ and $\alpha^g = 0$ if $\alpha = 0$.

(i) $A^g = (A'A)^g A' = A'(AA')^g$.

(j) $A^g A$, AA^g, $I - A^g A$, and $I - AA^g$ are all symmetric idempotent.

Example 1.1 If A is symmetric idempotent, all four conditions of [1.2] are satisfied by $A^g = A$, i.e. A is its own unique g-inverse.

Example 1.2 If A is symmetric, it is reducible to leading diagonal canonical form

$$HAH' = \begin{bmatrix} \Lambda & 0 \\ 0 & 0 \end{bmatrix},$$

i.e.

$$A = H' \begin{bmatrix} \Lambda & 0 \\ 0 & 0 \end{bmatrix} H = H_1' \Lambda H_1, \qquad [1.9]$$

where Λ is a diagonal matrix of the nonzero latent roots of A, and H is orthogonal and partitioned conformably as

$$H' = [H_1' \quad H_2'].$$

It may be verified that for this particular case

$$A^g = H' \begin{bmatrix} \Lambda^{-1} & 0 \\ 0 & 0 \end{bmatrix} H = H_1' \Lambda^{-1} H_1.$$

1.3 One-condition generalized inverses

The existence of a g-inverse satisfying only condition (1) was mentioned by Baer (1952), whose idea was followed up by Sheffield (1958) in a paper on the solution of linear equations. Wilkinson (1958), who disclosed that the idea was suggested to him by A.T. James, used a g_1-inverse to solve singular linear equations. It was, however, Bjerhammar (1958) and Bose (1959) who independently first put this concept on a firm footing.

DEFINITION 1.2 A $k \times n$ matrix A^{g_1} is said to be a g_1-inverse of the $n \times k$ matrix A if

$$AA^{g_1}A = A. \qquad [1.10]$$

The name one-condition g-inverse and its abbreviation will be used instead of names such as pseudo-inverse (Sheffield), effective inverse (Wilkinson), general matrix inverse (Bjerhammar), conditional inverse (Bose), g-inverse (Rao, 1962; Rohde, 1964; Searle, 1966), and solution matrix (John, 1964).

A g_1-inverse is, of course, non-unique, and A^{g_1} as defined by [1.10] represents a set of g_1-inverses of A. However, A^{g_1} will most commonly denote any one member of the set, and only in one passage (§1.4) is a notational distinction necessary. Moreover, it will be understood that in the same formula or context A^{g_1} denotes the same g_1-inverse of A, although it may well be that repeated A^{g_1}'s in the same formula could be different g_1-inverses of A. When it is desired to emphasize this, symbols such as G_1, G_2 will be used, but most often the use of different A^{g_1}'s in the same formula would be pointless. In addition the following conventions will be adopted in connection with the equality sign: if B is such that $ABA = A$, we may write $B = A^{g_1}$ (meaning that B is a g_1-inverse of A), or, as is sometimes more convenient, $A^{g_1} = B$ (meaning that a g_1-inverse of A is given by the matrix B). However, if all g_1-inverses of A are of some *general form* C (say), then we shall write $C \equiv A^{g_1}$ (or $A^{g_1} \equiv C$), i.e. C represents the set of all g_1-inverses of A.

THEOREM 1.3 If P_1 and P_2 are nonsingular matrices, then $(P_1AP_2)^{g_1} \equiv P_2^{-1}A^{g_1}P_1^{-1}$ (Rohde, 1964, p. 34).

Proof: $P_1AP_2(P_2^{-1}A^{g_1}P_1^{-1})P_1AP_2 = P_1AA^{g_1}AP_2 = P_1AP_2$, i.e. $P_2^{-1}A^{g_1}P_1^{-1} = (P_1AP_2)^{g_1}$. To justify the identity relationship, i.e. to show that every g_1-inverse of P_1AP_2 has the form $P_2^{-1}A^{g_1}P_1^{-1}$ for some g_1-inverse of A, a result due to Bose (1959) is used. Let $P_1AP_2 = B$; then the relation $BB^{g_1}B = B$ implies that

$$P_1^{-1}BP_2^{-1}(P_2B^{g_1}P_1)P_1^{-1}BP_2^{-1} = P_1^{-1}BP_2^{-1}, \text{ or } A(P_2B^{g_1}P_1)A = A.$$

It follows that $P_2B^{g_1}P_1 = A^{g_1}$. Hence $B^{g_1} = P_2^{-1}A^{g_1}P_1^{-1}$ for some A^{g_1}.

Since the correspondence between B^{g_1} and A^{g_1} is (1, 1), we may write $B^{g_1} \equiv P_2^{-1}A^{g_1}P_1^{-1}$.

General form for A^{g_1}

The existence of A^{g_1} is guaranteed by Theorem 1.1. However, the following result, due to Bose, establishes the most general form of g-inverses which satisfy only [1.10].

THEOREM 1.4 If A has rank r, and P_1 and P_2 are nonsingular matrices such that

$$P_1AP_2 = N = \begin{bmatrix} I_r & 0 \\ 0 & 0 \end{bmatrix},$$

then a matrix G is a g_1-inverse of A if and only if it is expressible in the form $G = P_2N^{g_1}P_1$, where

$$N^{g_1} \equiv \begin{bmatrix} I_r & U \\ V & W \end{bmatrix}$$

for arbitrary U, V, and W.

Proof: Let

$$N^* = \begin{bmatrix} X & U \\ V & W \end{bmatrix};$$

then

$$NN^*N = \begin{bmatrix} X & 0 \\ 0 & 0 \end{bmatrix},$$

and it follows that a necessary and sufficient condition for $N^* \equiv N^{g_1}$ is $X = I_r$. Thus

$$N^{g_1} \equiv \begin{bmatrix} I_r & U \\ V & W \end{bmatrix},$$

where U, V, and W are arbitrary. It now follows from Theorem 1.3 that

$$A^{g_1} \equiv P_2N^{g_1}P_1. \qquad [1.11]$$

Example 1.3 If A is symmetric,

$$A^{g_1} \equiv H'\begin{bmatrix}\Lambda^{-1} & U \\ V & W\end{bmatrix}H,$$

where H and Λ are as in Example 1.2 and U, V, and W are arbitrary matrices.

1.4 Solution of linear equations

The importance of g-inverses in the solution of linear equations was mentioned in §1.1. This subject has been discussed by virtually all the leading writers on g-inverses, some of the earlier and most important papers being those of Penrose (1955), Sheffield (1958), Bjerhammar (1958), Bose (1959), and Rao (1962).

THEOREM 1.5 A necessary and sufficient condition for B to be in the column-space of A is $AA^{g_1}B = B$.

Proof: The sufficiency is obvious. Conversely, if $B = AK$ (say), then $AA^{g_1}B = AA^{g_1}AK = AK = B$.

Corollary Similarly, a necessary and sufficient condition for B to be in the row-space of A is $BA^{g_1}A = B$.

This result has also been presented by Rao (1967).

THEOREM 1.6 A necessary and sufficient condition for the equations $AXB = H$ to have a solution is

$$AA^{g_1}HB^{g_1}B = H, \quad [1.12]$$

the condition of consistency. If this condition is satisfied, the general solution to the equation is

$$X = A^{g_1}HB^{g_1} + Z - A^{g_1}AZBB^{g_1}, \quad [1.13]$$

where Z is arbitrary.

Proof: If [1.12] holds, $A^{g_1}HB^{g_1}$ is a solution to $AXB = H$. Conversely, if X satisfies $AXB = H$, then $H = AXB = AA^{g_1}AXBB^{g_1}B = AA^{g_1}HB^{g_1}B$.

If X is any solution and X_0 a particular solution, $A(X - X_0)B = 0$, i.e. $X^* = X - X_0$ is a solution to $AXB = 0$. If X^* is any solution to $AXB = 0$, and $X = X_0 + X^*$, $AXB = AX_0B + AX^*B = H$, i.e. X is a solution. This proves that $X_0 + X^*$ gives the general solution.

Now $A(Z - A^{g_1}AZBB^{g_1})B = 0$, and if $AXB = 0$ then $X = X - A^{g_1}AXBB^{g_1}$, i.e. $X_0 = Z - A^{g_1}AZBB^{g_1}$ represents all possible solutions to $AXB = 0$. Since $A^{g_1}HB^{g_1}$ is a particular solution to $AXB = H$, the general solution is given by [1.13].

The proof of Theorem 1.6 is due to Penrose, except that he had A^g and B^g in place of A^{g_1} and B^{g_1}. He did point out, however, that only condition (1) was required for his proof.

Corollary 1 The equations $AXA = A$ are consistent since $AA^{g_1}AA^{g_1}A = A$, i.e. [1.12] holds, and the general solution is

$$X = A^{g_1}AA^{g_1} + U - A^{g_1}AUAA^{g_1},$$

where U is arbitrary. Each A^{g_1} may be different, or they may all be the same, i.e.

$$X = GAG + U - GAUAG, \qquad [1.14]$$

where $G = A^{g_1}$, also represents the general solution.

It follows that the right-hand side of [1.14] is a representation for all g_1-inverses of A, i.e.

$$A^{g_1} \equiv GAG + U - GAUAG. \qquad [1.15]$$

This result was given by Bjerhammar, but in slightly different form.

Corollary 2 For equations $AX = H$, the consistency condition is $AA^{g_1}H = H$ and the general solution is

$$X = A^{g_1}H + (I - A^{g_1}A)Z. \qquad [1.16]$$

From Theorem 1.5 the consistency condition is equivalent to the requirement that H is in the column-space of A.

Corollary 3 For equations $XA = H$, the consistency condition is $HA^{g_1}A = H$ (cf. Theorem 1.5, Corollary) and the general solution is $X = HA^{g_1} + Z(I - AA^{g_1})$.

Corollary 4 For equations $A\mathbf{x} = \mathbf{h}$, the consistency requirement is $AA^{g_1}\mathbf{h} = \mathbf{h}$ and the general solution is

$$\mathbf{x} = A^{g_1}\mathbf{h} + (I - A^{g_1}A)\mathbf{z}, \qquad [1.17]$$

where $\mathbf{z}$ is arbitrary.

If A is $n \times k$ of rank r and its rows and columns are arranged so that the leading $r \times r$ submatrix A_{11} in

$$A = \begin{bmatrix} A_{11} & A_{12} \\ A_{21} & A_{22} \end{bmatrix} \qquad [1.18]$$

is nonsingular, it is readily verified, using $A_{22} = A_{21}A_{11}^{-1}A_{12}$, that

$$\begin{bmatrix} A_{11}^{-1} & 0 \\ 0 & 0 \end{bmatrix}$$

satisfies [1.10]. Hence the general solution may be written as

$$\mathbf{x} = \begin{bmatrix} A_{11}^{-1}\mathbf{h}_1 \\ 0 \end{bmatrix} + \begin{bmatrix} -A_{11}^{-1}A_{12} \\ I \end{bmatrix}\mathbf{x}_2, \tag{1.19}$$

where the partitioning of $\mathbf{x}$ and $\mathbf{h}$ conforms with that of A. This is the usual representation, cf. Aitken (1956, p. 70).

It is well known, and shown by Searle (1966, p. 150) for the representation [1.17], that the number of linearly independent solutions is $k - r + 1$.

Corollary 5 If X is any solution to the consistent equations $AX = H$, then KX is unique if and only if $K = KA^{g_1}A$, i.e. if and only if K is in the row-space of A (Theorem 1.5, Corollary). Similarly for equations $XA = H$, the condition is $K = AA^{g_1}K$, i.e. K in the column-space of A (Theorem 1.5).

Proof: From Corollary 2 we may write the general solution as $X = GH + (I - A^{g_1}A)Z$, where G is some particular g_1-inverse of A. Hence if $K = KA^{g_1}A$, $KX = KGH = KA^{g_1}AGAA^{g_1}H = KA^{g_1}H$, which shows the uniqueness. Conversely, if KX is unique, $KA^{g_1}H$ is unique (taking $Z = 0$ in [1.16]) and therefore $K(I - A^{g_1}A)Z$ is unique for arbitrary Z, i.e. $K(I - A^{g_1}A) = 0$.

Corollary 6 If $\mathbf{x}$ is any solution to the consistent equations $A\mathbf{x} = \mathbf{h}$, then $\mathbf{k}'\mathbf{x}$ is unique if and only if $\mathbf{k}' = \mathbf{k}'A^{g_1}A$.

THEOREM 1.7 A necessary and sufficient condition for $X^* = GH$ to be a solution to the consistent equations $AX = H$ is $G = A^{g_1}$.

Proof: Since the equations are consistent, X exists such that $AX = H$. Thus, if $G = A^{g_1}$, $AX^* = AGH = AGAX = AX = H$, i.e. X^* is a solution. Conversely, if X^* is a solution, $AGH = H$ for all consistent H,

$$\text{i.e. } AGAU = AU \quad \text{for arbitrary } U,$$

$$\text{i.e. } AGA = A, \text{ or } G = A^{g_1}.$$

(This proof follows that of Bose (1959) for the Corollary.)

Corollary A necessary and sufficient condition for $\mathbf{x}^* = G\mathbf{h}$ to be a solution to the consistent equations $A\mathbf{x} = \mathbf{h}$ is $G = A^{g_1}$.

Some authors actually base their definition of a g_1-inverse on its properties in the solution of linear equations. Thus Rao (1962) defines a g_1-inverse of A as a matrix A^{g_1} such that, for any $\mathbf{h}$ for which $A\mathbf{x} = \mathbf{h}$ is consistent, $\mathbf{x} = A^{g_1}\mathbf{h}$ is a solution. Theorem 1.7,

Corollary, demonstrates that this definition and Definition 1.2 are equivalent.

Bjerhammar (1958) based his definition of A^{g_1} on similar considerations. He also asserted that $X = \{A^{g_1}\}H$, where $\{A^{g_1}\}$ denotes differentially the set of all g_1-inverses of A, is the general solution to consistent equations $AX = H$. This would mean, from [1.16], that $\{A^{g_1}\}H$ was an alternative representation for $GH + (I - GA)Z$, where G is some particular g_1-inverse of A. However, from [1.14],

$$\begin{aligned}\{A^{g_1}\}H &= GAGH + UH - GAUAGH \\ &= GAGAX + UH - GAUAGAX \\ &= GH + (I - GA)UH,\end{aligned}$$

i.e. $\{A^{g_1}\}H$ is the subset of [1.16] for which Z is in the row-space of H. Chipman (personal communication) has pointed out that, for $H = 0$, $\{A^{g_1}\}H$ would give the null matrix as the general solution to $AX = 0$.

However, as shown by Urquhart (1969), for the case $A\mathbf{x} = \mathbf{h}$ ($\mathbf{h} \neq \mathbf{0}$), $\{A^{g_1}\}\mathbf{h}$ does provide an alternative representation for $G\mathbf{h} + (I - GA)\mathbf{z}$, since it is always possible to find U such that $U\mathbf{h} = \mathbf{z}$, e.g. $U = \mathbf{z}(\mathbf{h}'\mathbf{h})^{-1}\mathbf{h}'$. Urquhart states a preference for the former over the latter representation since "the 'arbitrary term' $(I - GA)\mathbf{z}$ can be somewhat of a nuisance". However, our experience in writing this monograph is that the arbitrary term has a convenient habit of vanishing (through premultiplication by A or by any matrix in the row-space of A).

1.5 Two-condition generalized inverses

As early as 1955, Rao made use of a g-inverse which satisfied the first two conditions of [1.2]. Rao, who called such a matrix a "pseudo-inverse", originally considered positive semidefinite matrices only, but in 1962 he produced a generalization to any matrix. Bjerhammar, independently and through a different approach, defined an equivalent type of g-inverse. Bjerhammar's results in this connection are discussed in Chapter 2. The third type of g-inverse introduced in this monograph is thus defined as follows:

DEFINITION 1.3 A g_2-inverse of an $n \times k$ matrix A of rank $r \leqslant \min(n, k)$ is defined as a $k \times n$ matrix A^{g_2} such that

$$\left.\begin{aligned} AA^{g_2}A &= A, \\ A^{g_2}AA^{g_2} &= A^{g_2}. \end{aligned}\right\} \qquad [1.20]$$

This type of g-inverse has also been called a "reciprocal inverse"

(Bjerhammar) and a "reflexive inverse" (Rohde, 1964, p. 13).

THEOREM 1.8 If P_1 and P_2 are nonsingular matrices, then $(P_1 A P_2)^{g_2} \equiv P_2^{-1} A^{g_2} P_1^{-1}$ (Rohde, p. 35).

Proof: By Theorem 1.3 the first condition of [1.20] is satisfied by $P_2^{-1} A^{g_2} P_1^{-1}$. Furthermore,

$$P_2^{-1} A^{g_2} P_1^{-1} (P_1 A P_2) P_2^{-1} A^{g_2} P_1^{-1} = P_2^{-1} A^{g_2} A A^{g_2} P_1^{-1} = P_2^{-1} A^{g_2} P_1^{-1}.$$

If we put $P_1 A P_2 = B$, the existence of a (1, 1) correspondence between B^{g_2} and A^{g_2} may be established in a similar fashion to that for B^{g_1} and A^{g_1}. The relations $BB^{g_2}B = B$ and $B^{g_2}BB^{g_2} = B^{g_2}$ imply that

$$P_1^{-1} B P_2^{-1} (P_2 B^{g_2} P_1) P_1^{-1} B P_2^{-1} = P_1^{-1} B P_2^{-1}$$

and

$$P_2 B^{g_2} P_1 (P_1^{-1} B P_2^{-1}) P_2 B^{g_2} P_1 = P_2 B^{g_2} P_1.$$

Thus

$$A(P_2 B^{g_2} P_1) A = A$$

and

$$(P_2 B^{g_2} P_1) A (P_2 B^{g_2} P_1) = P_2 B^{g_2} P_1.$$

It follows that $P_2 B^{g_2} P_1 = A^{g_2}$ and thus $B^{g_2} = P_2^{-1} A^{g_2} P_1^{-1}$ for some A^{g_2}.

General form for A^{g_2}

In view of Theorem 1.1, every matrix A has a g_2-inverse. However, in this section a general form for g-inverses which satisfy only conditions [1.20] will be developed.

The result of Theorem 1.8 suggests that this general form may be obtained in a manner similar to that used for A^{g_1}, i.e. by finding the condition that

$$N^{g_1} \equiv \begin{bmatrix} I_r & U \\ V & W \end{bmatrix}$$

should be a g_2-inverse of

$$N = \begin{bmatrix} I_r & 0 \\ 0 & 0 \end{bmatrix}.$$

Since

$$N^{g_1} N N^{g_1} = \begin{bmatrix} I & U \\ V & W \end{bmatrix} \begin{bmatrix} I & 0 \\ 0 & 0 \end{bmatrix} \begin{bmatrix} I & U \\ V & W \end{bmatrix} = \begin{bmatrix} I & U \\ V & VU \end{bmatrix},$$

a necessary and sufficient condition for $N^{g_1} = N^{g_2}$ is $W = VU$. It now follows that

$$A^{g_2} \equiv P_2 N^{g_2} P_1 \qquad [1.21]$$

with

$$N^{g_2} \equiv \begin{bmatrix} I & U \\ V & VU \end{bmatrix}.$$

This proves

THEOREM 1.9 If P_1 and P_2 are nonsingular matrices such that

$$P_1 A P_2 = N = \begin{bmatrix} I_r & 0 \\ 0 & 0 \end{bmatrix},$$

then a matrix G is a g_2-inverse of A if and only if it can be expressed as $G = P_2 N^{g_2} P_1$, where

$$N^{g_2} \equiv \begin{bmatrix} I_r & U \\ V & VU \end{bmatrix},$$

for arbitrary U and V.

Rao established the existence of A^{g_2} by considering the reduction

$$P_1 A P_2 = \begin{bmatrix} D & 0 \\ 0 & 0 \end{bmatrix},$$

where P_1 and P_2 are nonsingular matrices and D is an $r \times r$ diagonal matrix of rank r. This reduction has also been outlined by Searle (1966, pp. 127–9). Clearly

$$A^{g_2} = P_2 \begin{bmatrix} D^{-1} & 0 \\ 0 & 0 \end{bmatrix}' P_1. \qquad [1.22]$$

It is curious to note that Searle (pp. 145–7) actually defines

$$P_2 \begin{bmatrix} D^{-1} & 0 \\ 0 & 0 \end{bmatrix}' P_1$$

as a g_1-inverse of A. Although from the inclusive viewpoint this is not incorrect, it is nevertheless confusing, as Searle actually appeals to the reflexive property of this form of g-inverse at later stages in his book.

An alternative form for g_2-inverses was used by Lucas (1962) who considered A partitioned as in [1.18]. It is easily verified that

$$\begin{bmatrix} A_{11}^{-1} & 0 \\ 0 & 0 \end{bmatrix}$$

also satisfies condition (2) and is therefore actually a g_2-inverse of A (cf. §1.4). In general it will be necessary to apply permutation matrices to A to produce a leading submatrix of rank r, but since such matrices are nonsingular, Theorem 1.8 may be applied to obtain A^{g_2}.

In deriving [1.19] we used a g_2-inverse for G, and, since a g_2-inverse is also a g_1-inverse, this was permissible.

Further properties of A^{g_2}, in particular the relationship between A^{g_1} and A^{g_2}, will be discussed in Chapter 2.

1.6 Three-condition generalized inverses

Goldman & Zelen (1964) proposed a type of g-inverse which they called a "weak generalized inverse". These authors required their g-inverse to satisfy conditions (1), (2), and (4). As was mentioned in the introduction, traditional statistical notation makes it convenient to replace condition (4) by condition (3). We therefore define a three-condition generalized inverse of A as follows:

DEFINITION 1.4 If A is an $n \times k$ matrix, then a g_3-inverse of A is a $k \times n$ matrix A^{g_3} satisfying

$$\left.\begin{aligned} AA^{g_3}A &= A, \\ A^{g_3}AA^{g_3} &= A^{g_3}, \\ (AA^{g_3})' &= AA^{g_3}. \end{aligned}\right\} \qquad [1.23]$$

(For Goldman & Zelen's g_3-inverse, denoted $A^{g_3^*}$, the last condition is replaced by $(A^{g_3^*}A)' = A^{g_3^*}A$). Rohde (p. 14) has suggested that A^{g_3} be called a "normalized generalized inverse".

We will show in Chapter 2 that every g_3-inverse of A is expressible as $(A'A)^{g_1}A'$. Goldman & Zelen have proved a similar result.

1.7 Discussion

The first point which will be discussed here is that of the inclusiveness of the four definitions of g-inverses presented in this chapter. Rohde (p. 96) has pointed out that for these four types of g-inverses we have the relationship

$$A^{g_1} \supseteq A^{g_2} \supseteq A^{g_3} \supseteq A^{g}, \qquad [1.24]$$

with equality holding throughout if and only if A is nonsingular. Clearly then, every g-inverse is at least a g_1-inverse. In view of this relationship it is necessary to use the definitions of the different types of g-inverses in an inclusive sense. However, at certain stages – for example, when discussing specific properties of a particular type of g-inverse, or when presenting formulae for computing the different types of g-inverses – it is necessary to discuss each type of g-inverse exclusively. It should be clear from the context exactly which conditions govern the type of g-inverse under discussion. Finally, as regards terminology, it will be convenient in certain cases to refer to A^{g_3} and A^g as higher order g-inverses of A.

The key condition in the definition of each type of g-inverse is condition (1). This condition can be used in conjunction with conditions (3) and/or (4) to produce further types of g-inverses. Certain of these alternative combinations possess interesting properties as regards the ability of a g-inverse to commute, and also in least squares estimation theory. These special cases will be briefly mentioned in later chapters.

Some attention will now be paid to aspects which will not be fully discussed in this monograph. First of all it is clear that nearly all the results presented so far hold when the matrices are permitted to have elements in the field of complex numbers, provided the transpose operation is interpreted as the conjugate transpose operation and symmetric is replaced by Hermitian (see Penrose, 1955; Bjerhammar, 1958; Price, 1964).

The study of the concept of g-inverses has not been limited to finite spaces. In this respect, it is recorded that a generalization to Hilbert space has been achieved by Desoer & Whalen (1963) and by Ben-Israel & Charnes (1963).

It will have been noticed that A^gA is the operator which projects onto the column-space of A'. Several authors (e.g. Greville, 1962; Albert & Sittler, 1966) do in fact study g-inverses in a vector-space framework. An analytic approach using the more familiar matrix techniques has been adopted here in preference to the vector-space concepts, at the expense of some inconvenience at stages where the geometrical approach is the simpler one.

Chapter 2

THEORETICAL PROPERTIES OF GENERALIZED INVERSES

2.1 Introduction

In the present chapter the algebraic implications of the four definitions of g-inverses presented in Chapter 1 will be considered in more detail. Three properties, namely rank, symmetry, and latent roots and vectors, are of primary importance in the characterization of any class of matrices. If A is a nonsingular matrix, there exists a simple and well-known relationship between each of these properties and the corresponding property of A^{-1}. It is natural therefore to determine whether analogous relationships hold for generalized inverses of A. The discussion of latent roots and vectors is, however, governed to some extent by the property of commutativity of A and a g-inverse of A. Some results on commuting g-inverses are therefore presented. The conditions under which the "reverse order law" related to ordinary inverses of matrix products, i.e. $(AB)^{-1} = B^{-1}A^{-1}$, is transferable to g-inverses forms the subject-matter of § 2.7. We also present results on the relationships between the different types of g-inverses, while the final two sections consist of a collection of useful results on g_1-inverses and on g-inverses in general, respectively.

2.2 Results on rank and idempotency

The material of this section is well known and has been presented by many authors.

THEOREM 2.1 Let A be $n \times k$; then

(a) the matrices AA^{g_1}, $A^{g_1}A$, $I - AA^{g_1}$, $I - A^{g_1}A$ are all idempotent with ranks equal to $\mathrm{r}(A)$, $\mathrm{r}(A)$, $n - \mathrm{r}(A)$, $k - \mathrm{r}(A)$ respectively,

(b) $\mathrm{r}(A^{g_1}) \geqslant \mathrm{r}(A)$.

Proof: Idempotency is easily proved using [1.10]. Since the rank of the product of two matrices does not exceed the rank of either factor, $\mathrm{r}(A) = \mathrm{r}(AA^{g_1}A) \leqslant \mathrm{r}(AA^{g_1}) \leqslant \mathrm{r}(A)$, i.e. $\mathrm{r}(AA^{g_1}) = \mathrm{r}(A)$. Similarly, $\mathrm{r}(A^{g_1}A) = \mathrm{r}(A)$.

To prove that $\mathrm{r}(I - AA^{g_1}) = n - \mathrm{r}(A)$, for example, we use the result that the rank and trace of an idempotent matrix are equal. Thus

$$\mathrm{r}(I - AA^{g_1}) = \mathrm{tr}(I - AA^{g_1}) = \mathrm{tr}(I) - \mathrm{tr}(AA^{g_1}) = n - \mathrm{r}(AA^{g_1}).$$

The result $r(A^{g_1}) \geqslant r(A)$ follows since $r(A^{g_1}A) = r(A)$, i.e. $r(A^{g_1}) \geqslant r(A^{g_1}A) = r(A)$.

Rao (1965) has shown that it is always possible to construct a g_1-inverse of an $n \times k$ matrix A, of rank $r \leqslant \min(n, k)$, which has rank equal to $\min(n, k)$. By [1.10],

$$r(A^{g_1}) = r(P_2 N^{g_1} P_1) = r(N^{g_1}) = r\begin{bmatrix} I_r & U \\ V & W \end{bmatrix};$$

hence U, V, and W, which are arbitrary matrices, may be selected so that N^{g_1} has independent rows (columns). This implies that for the case $n = k$ a nonsingular g_1-inverse of A exists. For example, if W in Example 1.3 is nonsingular, A^{g_1} has full rank.

It will be proved in Theorem 2.12 that the matrix $A(A'A)^{g_1}A'$ is idempotent, with rank equal to $r(A)$.

Bjerhammar pointed out that rank is the only criterion which distinguishes between A^{g_1} and A^{g_2}. The following proof of his result is offered:

THEOREM 2.2 A necessary and sufficient condition that A^{g_1} should be a g_2-inverse is $r(A^{g_1}) = r(A)$.

Proof: By Theorem 2.1, the condition $AGA = A$ implies $r(G) \geqslant r(A)$. Similarly, if $GAG = G$, then $r(A) \geqslant r(G)$. Thus $r(A^{g_2}) = r(A)$. This proves the necessity. As regards the sufficiency, if $r(A^{g_1}) = r(A) = r$, then from [1.10]

$$r(N^{g_1}) = r\begin{bmatrix} I_r & U \\ V & W \end{bmatrix} = r.$$

A necessary and sufficient condition for this to hold is $W = VU$. But by [1.21] this gives $N^{g_1} = N^{g_2}$ or $A^{g_1} = A^{g_2}$.

Corollary $r(A^{g_3}) = r(A^{g}) = r(A)$.

2.3 Results on symmetry

It is well known that, if a nonsingular matrix A is symmetric, A^{-1} is also symmetric. The following theorem indicates that the same result holds for the g-inverse of a symmetric matrix.

THEOREM 2.3 If A is symmetric then A^{g} is symmetric (Rohde, 1964, p. 37).

Proof: The proof follows immediately from Theorem 1.2(b), i.e. $A^g = (A')^g = (A^g)'$.

The representation for A^g given in Example 1.2 typifies the above result.

It is evident from the general forms for A^{g_1} and A^{g_2} that symmetry is not guaranteed when A is symmetric. However, it is easily seen from Example 1.3 that, if $V = U'$, and W is symmetrical, then A^{g_1} is symmetric.

Lucas (1962) establishes the existence of symmetric g_1-inverses in the following way: if $G = A^{g_1}$ and A is symmetric, then $G' = A^{g_1}$, so that $\frac{1}{2}(G + G')$ is a symmetric g_1-inverse.

The existence of a symmetric g_2-inverse also follows from Example 1.3, since, by [1.21], the symmetric matrix

$$H' \begin{bmatrix} \Lambda^{-1} & U \\ U' & U'U \end{bmatrix} H$$

is a g_2-inverse of A. It is also evident from the defining relation [1.20] that, if G is a g_2-inverse of the symmetric matrix A, then so is G'.

Rohde (1964, p. 36) states the following result for g_3-inverses:

THEOREM 2.4 If A and A^{g_3} are symmetric, then $A^{g_3} = A^g$.

Proof: Since A^{g_3} is symmetric

$$(A^{g_3}A)' = AA^{g_3} = (AA^{g_3})' = A^{g_3}A,$$

and the fourth condition is satisfied.

An existence result for g_3-inverses of symmetric matrices is thus of little importance, since A^g is necessarily symmetric if A is symmetric. We note also that for A symmetric $[(A^{g_3})'A]' = AA^{g_3} = (AA^{g_3})' = (A^{g_3})'A$, i.e. $(A^{g_3})' = A^{g_3^*}$.

For a *general* matrix A, it was shown in Theorem 1.2(b) that $(A^g)' = (A')^g$. Although the corresponding result in terms of g_1-inverses, namely $(A^{g_1})' \equiv (A')^{g_1}$, does not hold in general, transposition of $AGA = A$ shows that G' is one choice of $(A')^{g_1}$. A similar result holds for g_2-inverses. It can also be seen that for $G = A^{g_3}$, $G' = (A')^{g_3^*}$.

2.4 Commuting generalized inverses

In this section a brief description of the properties of commuting g-inverses will be presented. A g_1-inverse which satisfies the relationship $AA^{g_1} = A^{g_1}A$, i.e. a g_1-inverse which commutes with A, will be denoted by A^{cg_1}. (A necessary condition for a matrix to commute with A is, of course, that A be square.)

Englefield (1966) has established that a necessary and sufficient condition for the existence of A^{cg_1} is $r(A) = n - m_0(A)$, where n is the order of A and $m_0(A)$ denotes the multiplicity of the zero latent root of A. However, since this condition is only one of several equivalent conditions, Englefield's result may be restated and proved as follows:

THEOREM 2.5 If $r(A) = r \leqslant n$, then any one of the following conditions is necessary and sufficient for the existence of A^{cg_1}:

(a) $r(A) = n - m_0(A)$,

(b) $r(A) = r(A^2)$,

(c) there exists B nonsingular such that

$$BAB^{-1} = \begin{bmatrix} A_1 & 0 \\ 0 & 0 \end{bmatrix},$$

where A_1 is $r \times r$ nonsingular.

Proof: The equivalence of these three conditions has been established by Mirsky (1955, p. 273). Now, if A^{cg_1} exists, $A = AA^{cg_1}A = A^{cg_1}A^2$. This implies $r(A) \leqslant r(A^2)$. However, since at the same time $r(A^2) \leqslant r(A)$, only the equality holds, and so $r(A) = r(A^2)$ is a necessary condition. The sufficiency is easily proved from condition (c), for

$$B^{-1}\begin{bmatrix} A_1^{-1} & 0 \\ 0 & Z \end{bmatrix}B, \qquad [2.1]$$

where Z is an arbitrary submatrix, is a commuting g_1-inverse of A. (This shows that $r(A^{cg_1}) \geqslant r(A)$.)

Mitra (1968b) has also studied commuting g-inverses by investigating the structure of a g-inverse with row- and column-spaces belonging to specified row- and column-spaces. He also demonstrates the relationships which exist between this type of g-inverse and that defined by Scroggs & Odell (1966).

Corollary 1 If A is symmetric, A^{cg_1} always exists, and it is also possible to construct a symmetric commuting g_1-inverse of A.

Proof: The first part follows, since $\mathrm{r}(A^2) = \mathrm{r}(A'A) = \mathrm{r}(A)$. As regards the second part, it follows from [1.9] and [2.1] that

$$H' \begin{bmatrix} \Lambda^{-1} & 0 \\ 0 & W \end{bmatrix} H, \qquad [2.2]$$

where W is an arbitrary symmetric submatrix, is a symmetric commuting g_1-inverse.

Corollary 2 If A^{cg_1} exists, then so does a commuting g_2-inverse (A^{cg_2}).

Proof: If A^{cg_1} exists, A may be expressed as in Theorem 2.5 (c). Clearly,

$$B^{-1} \begin{bmatrix} A_1^{-1} & 0 \\ 0 & 0 \end{bmatrix} B = A^{cg_2}.$$

Example 2.1 From [2.2], if A is symmetric,

$$H' \begin{bmatrix} \Lambda^{-1} & 0 \\ 0 & 0 \end{bmatrix} H = A^{cg_2}.$$

Commuting g-inverses have many interesting properties. For example, Englefield shows that if G_1 and G_2 are two commuting g_1-inverses of A, then

$$AG_1 = G_2A. \qquad [2.3]$$

This follows since

$$\underline{AG_1} = AG_2\underline{AG_1} = \underline{AG_2G_1A} = G_2\underline{AG_1A} = G_2A.$$

Furthermore, for $A = A'$, the class of symmetric commuting g_1-inverses of A belongs to a new class of g-inverses. If $G = A^{cg_1}$ and $G = G'$, then

$$\left.\begin{aligned} AGA &= A \\ (AG)' &= GA = AG \\ (GA)' &= AG = GA, \end{aligned}\right\} \qquad [2.4]$$

i.e. symmetric commuting g_1-inverses satisfy conditions (1), (3), and (4). This may be readily verified with the matrix [2.2].

The property of commutativity of A and A^{cg_2} also serves to characterize a new type of unique g-inverse.

THEOREM 2.6 A commuting g_2-inverse is unique (Englefield).

Proof: Let G_1 and G_2 be two commuting g_2-inverses of A; then

$$G_1 = G_1\underline{AG_1} = G_1G_1\underline{A} = G_1G_1\underline{AG_2A} = G_1G_1A\underline{A}G_2 = G_1\underline{G_1A}A\underline{G_2A}G_2$$
$$= \underline{G_1AG_1}AAG_2G_2 = \underline{AG_2}G_2 = G_2.$$

Now Rohde (p. 45) has proved – and it is evident from [2.4] – that, if A and A^{cg_2} are symmetric, then $A^{cg_2} = A^g$. However, if A is symmetric, it follows from Theorem 2.6 that $A^{cg_2} = (A^{cg_2})'$. This gives a somewhat stronger result than Rohde's, namely:

Corollary If A is symmetric, then $A^{cg_2} = A^g$.

Clearly, then, a sufficient condition for A and A^g to commute is that A should be symmetric.(*) Penrose (1955) established this result for the more general class of normal matrices ($AA' = A'A$) and went on to show that the property of commutativity implied that $(A^n)^g = (A^g)^n$. However, the latter result is a special case of the following theorem:

THEOREM 2.7 If A is any matrix, then $(A^{cg_1})^n = (A^n)^{cg_1}$, where n is any positive integer.

Proof: The theorem is true for $n = 1$. Suppose it is true for some positive integer $k < n$. Then

$$\begin{aligned} A^{k+1}(A^{cg_1})^{k+1} &= A^k\underline{AA^{cg_1}}(A^{cg_1})^k \\ &= \underline{A^kA^{cg_1}A}(A^{cg_1})^k \\ &= A^k(A^{cg_1})^k \\ &= (A^{cg_1})^kA^k \\ &= (A^{cg_1})^kA^kA^{cg_1}A \\ &= (A^{cg_1})^{k+1}A^{k+1}. \end{aligned}$$

It is also apparent that $A^{k+1}(A^{cg_1})^{k+1} = AA^{cg_1} = A^{cg_1}A = (A^{cg_1})^{k+1}A^{k+1}$. Hence

$$A^{k+1}(A^{cg_1})^{k+1}A^{k+1} = AA^{cg_1}A^{k+1} = A^{k+1},$$

and the result follows by induction.

Corollary $A^{g_1}(A^{cg_1})^{n-1} = (A^n)^{g_1}$.

The proof follows the same lines as that of the main result.

(*) Pearl (1966) may be consulted for alternative conditions governing commutativity of A and A^g.

2.5 Latent roots and vectors

If A is a nonsingular matrix, the latent roots of A^{-1} are reciprocals of the latent roots of A and the corresponding latent vectors are the same. In this section the relationships between the latent roots and vectors of a square matrix and those of an associated g-inverse will be discussed.

THEOREM 2.8 Let G be a g-inverse of A, and λ a nonzero latent root of A with corresponding latent vector $\mathbf{x} \neq \mathbf{0}$. Then a sufficient condition for $G\mathbf{x} = \lambda^{-1}\mathbf{x}$ is $AG = GA$.

Proof: If $A\mathbf{x} = \lambda\mathbf{x}$, then $AGA\mathbf{x} = \lambda AG\mathbf{x}$, or $A\mathbf{x} = \lambda AG\mathbf{x}$, or $\mathbf{x} = AG\mathbf{x}$. If $AG = GA$, then $\mathbf{x} = GA\mathbf{x} = \lambda G\mathbf{x}$, i.e. $G\mathbf{x} = \lambda^{-1}\mathbf{x}$.

This result was pointed out by Englefield (1966), who also mentioned that since $\mathrm{r}(A^{cg_1}) \geqslant \mathrm{r}(A) = r$, A^{cg_1} will have r latent roots λ_i^{-1} and the remaining ones arbitrary. The latent vectors corresponding to these arbitrary latent roots belong to the null space of A.

Rao (1967) has derived an alternative sufficient condition for the result $G\mathbf{x} = \lambda^{-1}\mathbf{x}$, where G is some g-inverse of A, to hold. He shows that, if

$$G\sum_{i=1}^{t} a_i A^i = \sum_{i=1}^{t} a_i A^{i-1}$$

for some t and constants $a_1, a_2, \ldots, a_t$, then $A\mathbf{x} = \lambda\mathbf{x}$ implies $G\mathbf{x} = \lambda^{-1}\mathbf{x}$, provided that $\Sigma a_i \lambda^i \neq 0$.

In view of Theorem 2.6, Corollary, it is obvious that, for A symmetric, the nonzero latent roots of A^g are reciprocals of the nonzero latent roots of A, and the corresponding latent vectors of A and A^g are the same. This result was established by Rohde (1964, p. 45) for the class of normal matrices. If A is not symmetric then A^g need not possess either of the properties enjoyed by the unique symmetric g-inverse. For example, consider

$$A = \begin{bmatrix} 2 & 0 & 0 \\ 1 & 1 & 0 \\ 3 & 1 & 0 \end{bmatrix},$$

and

$$A^g = \frac{1}{6}\begin{bmatrix} 2 & -1 & 1 \\ -4 & 5 & 1 \\ 0 & 0 & 0 \end{bmatrix},$$

where the nonzero latent roots of A are 1 and 2 and those of A^g are 1 and $\frac{1}{6}$. Incidentally this example indicates that the g-inverse of a triangular matrix is not in general triangular.

Rohde (p. 39) has proved the following result for latent roots of g_3-inverses.

THEOREM 2.9 If A is symmetric, the nonzero latent roots of A and A^{g_3} are reciprocals.

Proof: If $A\mathbf{x} = \lambda\mathbf{x}$, then premultiplication of this equation by AA^{g_3} gives $\mathbf{x} = AA^{g_3}\mathbf{x} = (A^{g_3})'A\mathbf{x}$ or $(A^{g_3})'\mathbf{x} = \lambda^{-1}\mathbf{x}$. The result follows since a matrix and its transpose have the same latent roots.

Even if A is symmetric, no general relationships may be derived for A^{g_1} and A^{g_2}. This is easily verified by consideration of the matrix

$$A = \begin{bmatrix} 1 & 2 \\ 2 & 4 \end{bmatrix},$$

which has latent roots 0 and 5, and of

$$A^{g_1} = \begin{bmatrix} 1 & 0 \\ -2 & 1 \end{bmatrix} \quad \text{and} \quad A^{g_2} = \begin{bmatrix} 1 & 0 \\ 0 & 0 \end{bmatrix}$$

which have latent roots 1 and 1, and 0 and 1 respectively.

2.6 Relationships between types of generalized inverses

In this section we prove a series of results which show that g_2-inverses and higher order g-inverses are all expressible in terms of g_1-inverses.

g_2-inverses in terms of g_1-inverses

In Theorem 2.2 it was shown that the condition $r(A^{g_1}) = r(A)$ is necessary and sufficient for $A^{g_1} = A^{g_2}$. The following theorem expresses an alternative relationship between A^{g_1} and A^{g_2}:

THEOREM 2.10 A matrix G is a g_2-inverse of A if and only if it can be written as $G = G_1AG_2$, where G_1 and G_2 are any two g_1-inverses of A.

Proof: If $G = G_1AG_2$, then

$$AGA = A(G_1AG_2)A = A$$

and

$$GAG = (G_1AG_2)A(G_1AG_2) = G_1AG_2 = G.$$

As regards the necessity, by [1.11] and [1.21]

$$A^{g_1} \equiv P_2\begin{bmatrix} I & U \\ V & W \end{bmatrix}P_1 \quad \text{and} \quad A^{g_2} \equiv P_2\begin{bmatrix} I & U \\ V & VU \end{bmatrix}P_1,$$

whence

$$P_2\begin{bmatrix} I & U_2 \\ V_1 & V_1U_2 \end{bmatrix}P_1 = P_2\begin{bmatrix} I & U_1 \\ V_1 & W_1 \end{bmatrix}P_1P_1^{-1}\begin{bmatrix} I & 0 \\ 0 & 0 \end{bmatrix}P_2^{-1}P_2\begin{bmatrix} I & U_2 \\ V_2 & W_2 \end{bmatrix}P_1,$$

and the identity $G \equiv G_1AG_2$ holds. Equally $G \equiv G_1AG_1$.

The sufficiency was first noted by Bjerhammar (1958), who in fact introduced his "reciprocal inverse" as a matrix of the type G_1AG_2. Mitra (1968a) has also presented a proof of Theorem 2.10 but appears to have proved the sufficiency twice over without proving the necessity.

Corollary $AA^{g_2} = AA^{g_1}$ and $A^{g_2}A = A^{g_1}A$.

THEOREM 2.11 A matrix is a symmetric g_2-inverse of the symmetric matrix A if and only if it is expressible as GAG', where $G = A^{g_1}$.

Proof: Let G_1, $G_2 = A^{g_1}$; then by Theorem 2.10 $G_1AG_2 = A^{g_2}$. Now it was shown in § 2.2 that G_1' and G_2' are also g_1-inverses of A. Thus the matrix G_1AG_1' is a symmetric g_2-inverse. Conversely, if $A^{g_2} = (A^{g_2})'$, $G_1AG_2 = G_2'AG_1'$. Premultiplication by A gives $AG_2 = AG_1'$, i.e. symmetric g_2-inverses are of the form GAG'.

Corollary Every symmetric g_2-inverse of the matrix $S = A'A$ is positive semidefinite.

Proof: This follows since every symmetric g_2-inverse of S can be written as $(G_sA')(G_sA')'$, where $G_s = S^{g_1}$.

Theorem 2.10 also provides a simple proof of Theorem 2.6 and of its corollary.

g_3- and g_3^*-inverses in terms of g_1-inverses

Before proceeding to prove the main result, we prove the following theorem:

THEOREM 2.12 The matrix $A(A'A)^{g_1}A'$ is unique, symmetric idempotent, and has the same rank as A (Bose, 1959).

Proof: By definition, $A'A(A'A)^{g_1}A'A = A'A$. Now by Lemma 2.1 of Rayner & Livingstone (1965), which states that in any matrix equation it is possible to cancel A' as a premultiplier (postmultiplier) if it is followed (preceded) throughout by A, we obtain

$$A(A'A)^{g_1}A'A = A \qquad [2.5]$$

and

$$A'A(A'A)^{g_1}A' = A'. \qquad [2.6]$$

Let G_1 and G_2 be any two g_1-inverses of $A'A$. Then, by [2.5],

$$AG_1A'A = AG_2A'A,$$

i.e.

$$AG_1A' = AG_2A'.$$

This proves the uniqueness, which may be emphasized by writing

$$A(A'A)^{g_1}A' = A(A'A)^{g}A' = AA^{g} \qquad [2.7]$$

by Theorem 1.2 (i). Symmetry now follows from condition (3), and idempotency and rank from Theorem 2.1.

Corollary $A'(AA')^{g_1}A = A'(AA')^{g}A = A^{g}A$, a unique symmetric idempotent matrix of the same rank as A.

For convenience we reference separately here the versions of [2.5] and [2.6] in terms of AA', namely

$$AA'(AA')^{g_1}A = A, \qquad [2.8]$$

$$A'(AA')^{g_1}AA' = A'. \qquad [2.9]$$

Goldman & Zelen (1964) established that a matrix G is a g_3-inverse of A if and only if $G = G_sA'$, where G_s is any symmetric g_2-inverse of $S = A'A$. Rohde (1964, p. 44), on the other hand, established that a necessary and sufficient condition for G to be a g_3-inverse is $G = S^{g_1}A'$, where, of course, S^{g_1} depicts any g_1-inverse of S. It would appear at first sight that the latter condition represents a stronger result. However, it will be shown in the corollary to the following theorem that the two sets of conditions are equivalent, though Rohde's is simpler. We present Rohde's proof of his result.

THEOREM 2.13 A matrix G is a g_3-inverse of A if and only if it can be written as $G = (A'A)^{g_1}A'$.

Proof: If $G = (A'A)^{g_1}A'$, then since by [2.7] $AG = AA^{g}$, conditions (1), (2), and (3) are obviously satisfied, i.e. $G = A^{g_3}$. The necessity is established as in Goldman & Zelen. If $G = A^{g_3}$, then $G = G\underline{AG} = GG'A'$. It remains to show that $A^{g_3}(A^{g_3})' = (A'A)^{g_1}$. This follows since $A'AA^{g_3}\underline{(A^{g_3})'A'}A = A'AA^{g_3}AA^{g_3}A = A'A$. Thus

$$A^{g_3} \equiv (A'A)^{g_1}A'. \qquad [2.10]$$

Corollary 1 Let R_1 and R_2 denote the conditions

$$R_1: GA' \equiv A^{g_3} \iff G = (A'A)^{g_2} \text{ and } G = G',$$

$$R_2: GA' \equiv A^{g_3} \iff G = (A'A)^{g_1};$$

then R_1 and R_2 are equivalent.

Proof: Clearly R_2 implies R_1. Now by Theorem 2.11, if G is as in R_1, $G \equiv G_s A'AG_s'$, where $G_s = (A'A)^{g_1}$. Therefore

$$GA' \equiv G_s A'AG_s'A' \equiv G_s A' \quad \text{by [2.5]},$$

i.e. R_1 implies R_2.

Corollary 2 The matrix AA^{g_3} is unique, symmetric idempotent, and has the same rank as A (by Theorem 2.12).

Corollary 3 A matrix G is a g_3^*-inverse of A if and only if it is expressible as $G = A'(AA')^{g_1}$, i.e.

$$A^{g_3^*} \equiv A'(AA')^{g_1}, \qquad [2.11]$$

and $A^{g_3^*}A$ has the same properties as AA^{g_3} in Corollary 2.

Corollary 4 $A'AA^{g_3} = A'$, $A^{g_3^*}AA' = A'$.

These follow from [2.6] and [2.9]. We may note that they constitute a generalization of Theorem 1.2(d).

Properties of matrices of the form $(A'A)^{g_1}A'$ and $A'(AA')^{g_1}$ were first studied by Bjerhammar (1958). For example, he showed that $(A'A)^{g_1}A'$, which he called a "transnormal inverse", satisfies conditions (1) and (2), and that $A^{g_3} = A^{g_3}(A^{g_3})'A'$. However, Bjerhammar did not establish the symmetry of $A(A'A)^{g_1}A'$ in a general manner. Bjerhammar's name, "normal inverse", for the matrix $A'(AA')^{g_1}$ is remarkably similar to that proposed by Rohde for A^{g_3}.

A discussion on left and right inverses provides a useful addendum to the relations between g_1- and g_3-inverses presented thus far.

Left and right inverses

A regular inverse of a matrix A exists if and only if A is square and nonsingular. For a rectangular matrix A of order $n \times k$, a matrix R such that $AR = I_n$ is known as a right inverse of A and exists if and only if $r(A) = n$.

To find a right inverse of A, a matrix of full row-rank, we need to solve $AR = I_n$. By Theorem 1.6, Corollary 2, the condition of consistency is $AA^{g_1} = I_n$. Now by Theorem 1.4, $A^{g_1} \equiv P_2 N^{g_1} P_1$, where $N = P_1 A P_2 = [I_n \quad 0]$, i.e.

$$A^{g_1} \equiv P_2 \begin{bmatrix} I_n \\ V \end{bmatrix} P_1,$$

where V is arbitrary. Hence

$$AA^{g_1} = P_1^{-1}N\begin{bmatrix} I_n \\ V \end{bmatrix}P_1 = I_n, \qquad [2.12]$$

i.e. the condition is satisfied and R exists. That A must have full row-rank, if R exists, such that $AR = I_n$, is obvious from the rank-product rule.

Similarly, a matrix L such that $LA = I_k$ exists if and only if A has full column-rank, and for such a matrix $A^{g_1}A = I_k$. Clearly no matrix can possess both a left and a right inverse unless it is square and nonsingular.

If R is a right inverse of A, $R = A^{g_3}$. Conversely, if $G = A^{g_3}$, $AG = I_n$ by [2.12]; thus $R = A^{g_3}$. But from [2.12] it is also apparent that any g_1-inverse of A is also a g_3-inverse, i.e. $R = A^{g_1}$.

By Theorem 1.6, Corollary 2, the set of all right inverses of A may be written as $R = G + (I - GA)Z$, where $G = A^{g_1}$ and Z is arbitrary. For this special case (A of full row-rank), therefore,

$$A^{g_1} \equiv A^{g_3} \equiv G + (I - GA)Z.$$

The choice $Z = A'(AA')^{-1}$ gives $R = A'(AA')^{-1}$, a particularly useful form of right inverse. Actually, from [2.11], $A'(AA')^{-1} = A^{g_3^*} = A^g$, since $(AA')^g = (AA')^{-1}$ (cf. [1.8]). Thus, when A has full row-rank, $A^{g_1} = A^{g_2} = A^{g_3}$ and $A^{g_3^*} = A^g$, so that equality is narrowly missed in [1.24].

If A has full column-rank, $A^{g_1} = A^{g_2} = A^{g_3^*}$ and $A^{g_3} = A^g = (A'A)^{-1}A'$, and these comprise the left inverses of A.

Some special cases in the solution of linear equations (§1.4) may now be dealt with. Consider the set of consistent equations $AX = H$. By Theorem 1.6, Corollary 5, $X(= IX)$ is a unique solution if and only if $A^{g_1}A = I$, and thus if and only if A has full column-rank. This is further demonstrated by considering the general solution for X (Theorem 1.6, Corollary 2), namely $X = A^{g_1}H + (I - A^{g_1}A)Z$. If A admits a left inverse, the term involving Z vanishes. Hence the general solution reduces to

$$\begin{aligned} A^{g_1}H &= A^{g_3^*}AA^{g_1}H \quad \text{(since the equations are consistent)} \\ &= A^{g}AA^{g_1}H \quad \text{(by Theorem 2.13, Corollary 3)} \\ &= A^{g}H. \end{aligned}$$

The same solution may be obtained by normalizing the equations to $A'AX = A'H$, whence $X = (A'A)^{-1}A'H = A^gH$. Of course, if $AX = H$ is not consistent, and A has full column-rank, $X = (A'A)^{-1}A'H$ is the familiar least-squares solution employed in statistics.

The above remarks obviously also apply to the case where H is a column vector, while similar results hold for equations $XA = H$, if A has full row-rank.

The g-inverse in terms of g_1-inverses

Two results are presented here, the first being an immediate consequence of [2.7] and Theorem 2.12, Corollary 1.

THEOREM 2.14 $A^g = A'(AA')^{g_1}A(A'A)^{g_1}A'$ (Bjerhammar, 1958).

Corollary By Theorem 2.13 and its Corollary 3 we may write $A^g = A^{g_3^*}AA^{g_3}$, or, if A is symmetrical, $A^g = (A^{g_3})'AA^{g_3}$.

THEOREM 2.15 If G is a symmetric g_1-inverse of $S = A'A$, then a necessary and sufficient condition for $GA' = A^g$ is that G and S should commute.

Proof: Theorem 2.13 established that $S^{g_1}A' = A^{g_3}$. Therefore $GA' = A^g$ if and only if the fourth condition is satisfied, i.e. if and only if $(GS)' = GS$, i.e. if and only if $SG = GS$.

Miscellaneous relationships

We list here, without commenting on their importance, a number of additional relationships which have been noticed. The matrix S denotes $A'A$.

(a) $A^{cg_2} \equiv A^{cg_1}AA^{cg_1}$.

(b) $A^{g_2}(A^{g_3})' = S^{g_2}$, $A^{g_3}(A^{g_2})' = S^{g_2}$.

(c) $A^{g_3}(A^{g_3^*})' = A^{g_3}(A^g)' = S^{g_3}$ $(=(A^{g_3})^2$ if A is symmetric).

(d) $A^{g_3^*}(A^{g_3})' = A^g(A^{g_3})' = S^{g_3}$ $(=(A^{g_3^*})^2$ if A is symmetric).

(e) $(S^2)^{g_1}S = S^{g_1}SS^g = S^{g_1}S^gS = S^{g_3}$.

(f) $S^{g_1}S^g = (S^{g_3})^2 = (S^2)^{g_3}$.

(g) $(S^{g_3})^2 = S^{g_3}$.

To show (a) we use Theorem 2.10. That the relations (b) satisfy [1.20] with respect to S may be easily verified; they constitute a slight tightening up of Goldman & Zelen's (1964) result, $A^{g_3}(A^{g_3})' = S^{g_2}$. Direct verification assisted by (b) and Theorem 2.14, Corollary, proves (c), the last part of which follows since, for A symmetric, $(A^{g_3^*})' = A^{g_3}$. The latter result also gives (d) from (c). Result (e) follows from Theorems 2.13 and 1.2(d), and may be used to verify (f), the last part of which follows from (c) for the symmetric case. Finally, (g) is the consequence of (f) and Theorem 2.13.

It is interesting to note that, from (e), $S^{g_3}A' = S^{g_1}SS^{g}A' = S^{g_1}A'$ by [2.6], and hence that $S^{g_1}A' = S^{g_2}A' = S^{g_3}A'$ (Theorem 2.13, Corollary 1), whereas $S^{g}A' = A^{g}$.

2.7 Generalized inverses of product matrices

The results of Theorems 1.3 and 1.8 permit the evaluation of g_1-inverses and g_2-inverses of triple products of the type P_1AP_2, where P_1 and P_2 are nonsingular matrices. A slight generalization of these two theorems is possible.

THEOREM 2.16 If P_1 has full column-rank and P_2 has full row-rank and A is any matrix, $(P_1AP_2)^{g_1} \equiv P_2^{g_3}A^{g_1}P_1^{g_3^*}$.

Proof: It was seen in the previous section that $P_1^{g_3^*}$ and $P_2^{g_3}$ are left and right inverses respectively of P_1 and P_2. Let $P_1AP_2 = B$; then the relation $AA^{g_1}A = A$ implies that

$$P_1AP_2(P_2^{g_3}A^{g_1}P_1^{g_3^*})AP_2 = P_1AP_2 \quad \text{or} \quad B(P_2^{g_3}A^{g_1}P_1^{g_3^*})B = B,$$

i.e. $P_2^{g_3}A^{g_1}P_1^{g_3^*}$ is a g_1-inverse of B, for some g_1-inverse of A. Furthermore $BP_2^{g_3}P_2 = P_1AP_2P_2^{g_3}P_2 = B$, and $P_1P_1^{g_3^*}B = P_1P_1^{g_3^*}P_1AP_2 = B$; hence the relation $BB^{g_1}B = B$ implies that

$$P_1^{g_3^*}BP_2^{g_3}(P_2B^{g_1}P_1)P_1^{g_3^*}BP_2^{g_3} = P_1^{g_3^*}BP_2^{g_3},$$

or, since $A = P_1^{g_3^*}BP_2^{g_3}$,

$$A(P_2B^{g_1}P_1)A = A.$$

Thus $P_2B^{g_1}P_1$ is a g_1-inverse of A, for some g_1-inverse of B. The correspondence between A^{g_1} and B^{g_1} is thus (1, 1) and the "identically equal" sign may be used.

Rao (1962) proved the above result for the case $P_1 = I$.

Corollary 1 If P_1 and P_2 are as in the theorem, then

$$(P_1AP_2)^{g_2} \equiv P_2^{g_3}A^{g_2}P_1^{g_3^*}.$$

The proof of this result follows the same lines as the above theorem and Theorem 1.8.

The results of Theorem 2.16 do not hold in general for the higher order g-inverses. However, weaker versions of the theorem are given in Corollaries 2 to 5.

Corollary 2 If P_2 is as in the theorem and U is any column-orthonormal matrix ($U'U = I$), then $(UAP_2)^{g_3} \equiv P_2^{g_3}A^{g_3}U'$.

Proof: By (2.10)

$$(UAP_2)^{g_3} \equiv [(UAP_2)'(UAP_2)]^{g_1}(UAP_2)'$$
$$\equiv (P_2' A'AP_2)^{g_1}P_2' A'U'.$$

Now, since P_2' and P_2 are matrices admitting left and right inverses $(P_2^{g_3})'$ and $P_2^{g_3}$ respectively, it follows from the theorem that $(P_2' A'AP_2)^{g_1} \equiv P_2^{g_3}(A'A)^{g_1}(P_2^{g_3})'$, whence $(P_2' A'AP_2)^{g_1}P_2' A'U' \equiv P_2^{g_3} A^{g_3} U'$.

Corollary 3 If P_1 is as in the theorem and V is any column-orthonormal matrix, $(P_1AV')^{g_3^*} = VA^{g_3^*}P_1^{g_3^*}$.

Corollary 4 If U and V are any column-orthonormal matrices, then $(UAV')^g = VA^gU'$ (Rohde p. 35).

This result is easily verified from the defining relations [1.2].

Corollary 5 If P_1 and P_2 are as in the theorem and

(a) A has full row-rank, then $(P_1AP_2)^{g_3} \equiv P_2^{g_3}A^{g_3}P_1^g$,

(b) A has full column-rank, then $(P_1AP_2)^{g_3^*} \equiv P_2^gA^{g_3^*}P_1^{g_3^*}$,

(c) A is nonsingular, then $(P_1AP_2)^g = P_2^gA^{-1}P_1^g$.

Proof: In (a), (b), and (c), the first two conditions hold by Corollary 2. Now consider

$$P_1AP_2P_2^{g_3}A^{g_3}P_1^{g_3^*} = P_1P_1^{g_3^*};$$

condition (3) will hold if and only if $P_1P_1^{g_3^*}$ is symmetrical, i.e. if and only if $P_1^{g_3^*} = P_1^g$. Similarly

$$P_2^{g_3}A^{g_3^*}P_1^{g_3^*}P_1AP_2 = P_2^{g_3}P_2,$$

so that condition (4) holds if and only if $P_2^{g_3} = P_2^g$. It is now apparent for A nonsingular that $P_2^gA^{-1}P_1^g$ satisfies all four conditions with respect to P_1AP_2.

If P_1 and P_2 are nonsingular matrices, such that $P_1AP_2 = B$, then in the light of Corollary 4 above, $P_2^{-1}A^gP_1^{-1}$ is not in general the g-inverse of B. However, this latter matrix is nevertheless a uniquely determined g_2-inverse of B for given P_1 and P_2. Chipman (1964, p. 1084) has discussed this aspect for the case where P_1 and P_2 are positive definite symmetric matrices. The matrix $P_2^{-1}A^gP_1^{-1}$ is similar to Chipman's $B^‡$. For the case $P_1 = I$ and $P_2 = W$, where W is positive definite symmetric, Greville (1961) has in a similar manner defined what he calls the "W-pseudo-inverse". In this case the matrix $W^{-1}A^g$ is a uniquely determined g_3-inverse of B.

Several papers have been devoted to a study of the conditions under which the reverse order law $(AB)^g = B^gA^g$ holds. In various particular cases this relationship certainly holds. For example, if $B = A'$, or if A or B' is column-orthonormal, or if A and B admit right and left inverses respectively (cf. [1.8]), then the relationship is true. However, it is evident from Corollary 4 of Theorem 2.16 that in general the relationship does not hold. We record here the papers of Cline (1964b), Erdelyi (1966), and Greville (1966), who have considered the problem in some detail.

An interesting case, with statistical applications, where the reverse order law does hold, is given by Lewis & Odell (1966). These authors point out that for the case A of order $n \times k$ $(n \leqslant k)$ and rank n, the g-inverse of $A'VA$, for nonsingular V, is given by $A^g V^{-1} (A^g)'$. This is a special case of Theorem 2.16, Corollary 5.

2.8 Special results on g_1-inverses

In this section we prove three theorems which find considerable use in statistical applications.

THEOREM 2.17 If L is a $q \times k$ matrix contained in the row-space of the $k \times k$ matrix $S = A'A$, then $LS^{g_1}L'$ is unique, has the same rank as L, and is positive semidefinite.

Proof: Since L may be written as BS,

$$LS^{g_1}L' = BSS^{g_1}SB' = BSS^{g}SB' \quad \text{by [2.7]}$$
$$= LS^{g}L',$$

i.e. $LS^{g_1}L'$ is unique. Further, $LS^{g_1}L' = BSB'$, i.e. $LS^{g_1}L'$ is positive semidefinite, and $r(LS^{g_1}L') = r(BSB') = r(BS) = r(L)$.

Clearly, Theorem 2.17 is a stronger version of Theorem 2.12.

Corollary 1 If $LS^{g_1}L' = Q$, then $L'Q^{g_1}Q = L'$ and $QQ^{g_1}L = L$.

Proof: Since Q is in the row-space of L' and column-space of L and has the same rank as L, it follows that the row-space of L' is the same as that of Q. The corollary now follows from Theorem 1.5 and its corollary.

Corollary 2 If $LS^{g_1}L' = Q$, then $L'Q^{g_1}L$ is unique, positive semidefinite, and has the same rank as L.

Proof: The corollary is a direct application of the theorem and Corollary 1.

Results similar to those given in this theorem and its corollaries have been presented by Mitra & Rao (1968a, b).

THEOREM 2.18 $(S + L'L)^{g_1} = S^{g_1} - S^{g_1}L'(I + LS^{g_1}L')^{-1}LS^{g_1}$.

Proof: From Theorem 2.17, $LS^{g_1}L' = LS^{g}L'$ is positive semidefinite, and thus $I + LS^{g}L'$ possesses a regular inverse. The proof consists of verifying the defining relation [1.10]:

$$\begin{aligned} & (S + L'L)\{S^{g_1} - S^{g_1}L'(I + LS^{g}L')^{-1}LS^{g_1}\} \\ = \; & SS^{g_1} + L'LS^{g_1} - L'(I + LS^{g}L')^{-1}LS^{g_1} - L'(LS^{g}L')(I + LS^{g}L')^{-1}LS^{g_1} \\ = \; & SS^{g_1} + L'LS^{g_1} - L'(I + LS^{g}L')(I + LS^{g}L')^{-1}LS^{g_1} \\ = \; & SS^{g_1}. \end{aligned}$$

Also $SS^{g_1}(S + L'L) = S + L'L$. In this reduction use has been made of the relations

$$LS^{g_1}S = L \quad \text{and} \quad SS^{g_1}L' = L' \qquad [2.13]$$

which hold by virtue of Theorem 1.5 and its corollary.

It is interesting to note that the result

$$S^{g} - S^{g}L'(I + LS^{g}L')^{-1}LS^{g} = (S + L'L)^{g}$$

also holds. Analogous expressions for $(S + L'L)^{g_2}$ and $(S + L'L)^{g_3}$ may also be obtained. These results represent restricted generalizations of the corresponding formula for the regular inverse of a matrix of the form $A + H'H$ as presented, for example, by Rao (1965, p. 29).

The following two theorems presented by Chipman (1964) are of particular importance in least-squares theory.

THEOREM 2.19 Let A and H be complementary matrices. Then

(a) $A(A'A + H'H)^{-1}H' = 0$;

(b) $(A'A + H'H)^{-1} = (A'A)^{g_1}, \quad (A'A + H'H)^{-1} = (H'H)^{g_1}$;

(c) $(A'A + H'H)^{-1}A' = A^{g_3}, \quad (A'A + H'H)^{-1}H' = H^{g_3}$.

The proof is basically that of Chipman.

Proof:

(a) Let A be $n \times k$ of rank r and H be $q \times k$ of rank p, where $r + p = k$. We may assume without loss of generality that the first r rows A_1 of A have rank r, so that A can be written as

$A = \begin{bmatrix} A_1 \\ A_2 \end{bmatrix} = \begin{bmatrix} I_r \\ N \end{bmatrix} A_1$. Similarly H can be written as $H = \begin{bmatrix} H_1 \\ H_2 \end{bmatrix} = \begin{bmatrix} I_p \\ M \end{bmatrix} H_1$,

where $r(H_1) = p$. Let $\begin{bmatrix} A_1 \\ H_1 \end{bmatrix}^{-1} = [X \quad Y]$. Then

$$\begin{bmatrix} A_1 \\ H_1 \end{bmatrix} [X \quad Y] = \begin{bmatrix} A_1X & A_1Y \\ H_1X & H_1Y \end{bmatrix} = \begin{bmatrix} I_r & 0 \\ 0 & I_p \end{bmatrix}.$$

To prove that $A(A'A + H'H)^{-1}H' = 0$, we note that

$$W = \begin{bmatrix} A \\ H \end{bmatrix}$$

has rank k, so that $W'W = A'A + H'H$ is positive definite and invertible. Since

$$AY = \begin{bmatrix} I_r \\ N \end{bmatrix} A_1Y = 0 \quad \text{and} \quad HY = \begin{bmatrix} I_p \\ M \end{bmatrix} H_1Y = \begin{bmatrix} I_p \\ M \end{bmatrix},$$

$W'WY = H'HY = H_1'[I_p \quad M'] \begin{bmatrix} I_p \\ M \end{bmatrix}$. Premultiplication by $A(W'W)^{-1}$ gives

$$AY = A(W'W)^{-1}H_1'[I_p \quad M'] \begin{bmatrix} I_p \\ M \end{bmatrix} = 0,$$

i.e. $$A(W'W)^{-1}H_1'[I_p \quad M'] = 0,$$

or $$A(W'W)^{-1}H' = 0.$$

(b) The relation $(W'W)^{-1}W'W = I$ implies that

$$(A'A + H'H)^{-1}A'A + (A'A + H'H)^{-1}H'H = I. \qquad [2.14]$$

On premultiplication by $A'A$ the second term vanishes by (a), so that $(A'A + H'H)^{-1} = (A'A)^{g_1}$. Similarly, premultiplication by $H'H$ gives $(A'A + H'H)^{-1} = (H'H)^{g_1}$.

(c) These results follow directly from (b) and [2.10].

Corollary If A, H, and W are as defined in Theorem 2.19, then $AH' = 0$ is a necessary and sufficient condition for $(W'W)^{-1}A' = A^g$ and $(W'W)^{-1}H' = H^g$.

Proof: By Theorem 2.15 a necessary and sufficient condition for $(W'W)^{-1}A' = A^g$ is that $(W'W)^{-1}$, a g_1-inverse of $A'A$, should commute with $A'A$, which is the same as that $W'W$ and $A'A$ should commute, namely $A'AH'H = H'HA'A$, and this is so if $AH' = 0$. Conversely, if $A'AH'H = H'HA'A$,

$$\begin{aligned} AH' &= A(W'W)^{-1}A'AH'H(W'W)^{-1}H' \quad \text{by [2.5] and [2.6]} \\ &= A(W'W)^{-1}H'HA'A(W'W)^{-1}H' \\ &= 0, \end{aligned}$$

by Theorem 2.19(a). Clearly $AH' = 0$ is also necessary and sufficient for $(W'W)^{-1}H' = H^g$.

It is interesting to note that the result $(W'W)^{-1}A' = A^g$, under the condition $AH' = 0$, can be verified directly from Theorem 1.2. By Theorem 1.2 (f) and (g), the condition $AH' = 0$ implies that $(W'W)^{-1} = (W'W)^g = A^g(A^g)' + H^g(H^g)'$. Now by Theorem 2.22(b) (see § 2.9), $AH' = 0$ implies $AH^g = 0$, so that $(W'W)^{-1}A' = A^g\underline{(A^g)'A'} + H^g\underline{(H^g)'A'} = A^gAA^g + H^g(AH^g)' = A^g$ (cf. Rohde, p. 59).

THEOREM 2.20 A matrix G is a g_1-inverse of A if and only if $GA = (A'A + H'H)^{-1}A'A$, where H is complementary to A.

Proof: Let A be $n \times k$ of rank r. If $GA = (A'A + H'H)^{-1}A'A$, then by Theorem 2.19(b) and [2.5] $AGA = A$, i.e. $G = A^{g_1}$.

Conversely, if $G = A^{g_1}$, $\text{r}(GA) = r$ by Theorem 2.1(a). Hence there exists a matrix H of rank $k - r$ such that $HGA = 0$. To show that H is complementary to A, let $\mathbf{u}$ and $\mathbf{v}$ be any vectors such that $\mathbf{u}'A + \mathbf{v}'H = 0$; then

$$\mathbf{u}'A = \mathbf{u}'AGA + \mathbf{v}'HGA = (\mathbf{u}'A + \mathbf{v}'H)GA = 0,$$

i.e. $\mathbf{u}'A = \mathbf{v}'H = 0$ and there is no dependence relation across the row-spaces of A and H. Since $\text{r}(A) + \text{r}(H) = k$, A and H are complementary. Furthermore

$$A'A = A'AGA + H'HGA = (A'A + H'H)GA = W'WGA,$$

and $(W'W)^{-1}$ exists, i.e.

$$GA = (A'A + H'H)^{-1}A'A.$$

2.9 Miscellaneous results

This section consists of an assorted collection of interesting, and in some cases useful, results on g-inverses.

THEOREM 2.21 The following three statements are equivalent:

(a) $A'B = 0$,

(b) $A^{g_3}B = 0$,

(c) $B^{g_3}A = 0$.

Proof: By Theorem 2.13, Corollary 4, $A'B = A'AA^{g_3}B$, and (b) implies (a). If $A'B = 0$, $(A'A)^{g_1}A'B = A^{g_3}B = 0$, i.e. (a) implies (b). Equivalence of (a) and (c), and hence of (b) and (c), follows by interchange of letters.

THEOREM 2.22

(a) $B^gA^g = 0$ if and only if $AB = 0$ (Cline, 1964a),

(b) $AB^g = 0$ if and only if $AB' = 0$.

Proof: If $AB = 0$, then transposing and pre- and postmultiplying by $(B'B)^g$ and $(A'A)^g$ gives, by Theorem 2.2 (i), $B^gA^g = 0$. Conversely, if $B^gA^g = 0$, then $B'BB^gA^gAA' = B'(B^g)'B'A'(A^g)'A' = B'A' = (AB)' = 0$. This proves (a). A proof of (b) can be constructed along similar lines.

Generalized inverses of direct products

It is interesting to note that the g-inverse of a matrix A, a direct product (Kronecker product) of matrices $A_1, A_2, \ldots, A_k$, is the direct product of the A_i^g taken in the same order (Greville, 1961). Thus, for example, $(A_1 \dot{\times} A_2)^g = A_1^g \dot{\times} A_2^g$, where the symbol $\dot{\times}$ denotes direct product. Clearly, this relationship also holds for the lower order g-inverses.

Chapter 3

GENERALIZED INVERSES OF PARTITIONED AND BORDERED MATRICES

3.1 Introduction

Partitioned and bordered matrices are widely used by statisticians as theoretical and computational aids. A study of g-inverses of such matrices is thus of considerable importance.

For special cases of partitioned matrices, g-inverses are obvious by inspection. For example,

$$\begin{bmatrix} A & 0 \\ 0 & B \end{bmatrix}^{g_1} = \begin{bmatrix} A^{g_1} & 0 \\ 0 & B^{g_1} \end{bmatrix}, \qquad [3.1]$$

and similar results hold for the other types of g-inverses of a quasi-diagonal matrix. For a matrix of the form

$$\begin{bmatrix} A & B \\ 0 & 0 \end{bmatrix}$$

and with the columns of B in the column-space of A,

$$\begin{bmatrix} A & B \\ 0 & 0 \end{bmatrix}^{g_1} = \begin{bmatrix} A^{g_1} & 0 \\ 0 & 0 \end{bmatrix},$$

as may be readily verified from Theorem 1.5; and if A is nonsingular,

$$\begin{bmatrix} A & B \\ 0 & 0 \end{bmatrix}^{g_3} = \begin{bmatrix} A^{-1} & 0 \\ 0 & 0 \end{bmatrix}.$$

However, with the exception of various special cases such as Lucas's g_2-inverse given at the end of §1.5, expressions for g-inverses of the general four-way partitioned matrix do not appear to be available, and the representations for g-inverses of a general partitioned matrix are limited to the case of a two-way partitioning. These expressions, which are presented in the next section, are obtained only for a column-wise partitioning, although it is evident that analogous results hold when a row-wise partitioning of the matrix is made. For positive semidefinite matrices partitioned into four submatrices, expressions for g-inverses do, however, exist. These representations are derived in § 3.3.

The rest of the chapter will be devoted to the development of expressions for g-inverses of a bordered matrix. These expressions are particularly applicable to the theory of constrained linear models.

3.2 Matrices partitioned into two submatrices

Let an $n \times k$ matrix A, of rank $r \leqslant \min(n, k)$, be partitioned column-wise as

$$A = [U \quad V], \tag{3.2}$$

where U is of order $n \times s$.

THEOREM 3.1 If A is partitioned as in [3.2], then

$$A^{g_1} = \begin{bmatrix} U^{g_1} - U^{g_1}VC^{g_1}(I - UU^{g_1}) \\ C^{g_1}(I - UU^{g_1}) \end{bmatrix}, \tag{3.3}$$

where $C = (I - UU^{g_1})V$.

Proof: Formula [3.3] can be directly verified by forming $AA^{g_1}A$ and simplifying. It is of some interest, however, to observe that [3.3] can be derived in a straightforward manner using the results of the previous chapters. Consider the product $B = P_1 A P_2$, where

$$P_1 = \begin{bmatrix} U^{g_1} \\ I_n \end{bmatrix} \quad \text{and} \quad P_2 = \begin{bmatrix} I_s & -U^{g_1}V \\ 0 & I_{k-s} \end{bmatrix}.$$

Then

$$B = \begin{bmatrix} U^{g_1}U & U^{g_1}C \\ U & C \end{bmatrix},$$

and by inspection we find that, since $UU^{g_1}C = 0$,

$$B^{g_1} = \begin{bmatrix} I & 0 \\ -C^{g_1}U & C^{g_1} \end{bmatrix}.$$

Since P_1 has full column-rank, it follows from Theorem 2.16 that

$$A^{g_1} = P_2 B^{g_1} P_1 = \begin{bmatrix} U^{g_1} - U^{g_1} V C^{g_1}(I - UU^{g_1}) \\ C^{g_1}(I - UU^{g_1}) \end{bmatrix}.$$

The reader is reminded that U^{g_1} stands for any g_1-inverse of U but the same one throughout; similarly for C^{g_1} (cf. §1.3).

THEOREM 3.2 If A is partitioned as in [3.2], then

$$A^{g_2} = \begin{bmatrix} U^{g_1} U U^{g_1} - U^{g_1} V C^{g_2}(I - UU^{g_1}) \\ C^{g_2}(I - UU^{g_1}) \end{bmatrix},$$

where C is as above.

Proof: It is readily verified that, if $P_1 A P_2 = B$, where P_1, P_2, and B are as in Theorem 3.1, then

$$B^{g_2} = \begin{bmatrix} U^{g_1} U & 0 \\ -C^{g_2} U & C^{g_2} \end{bmatrix},$$

so that $A^{g_2} = P_2 B^{g_2} P_1$ by Theorem 2.16, Corollary 1, and the result follows.

Other choices of B^{g_1} and B^{g_2} in the above theorems are possible, and it is feasible that certain of these choices might lead to simpler expressions for A^{g_1} and A^{g_2}. If $C = 0$, which, for example, is certainly true if V is in the column-space of U, then it is permissible to choose $C^{g_1} = C^{g_2} = 0$.

As regards Theorem 3.2, it must be observed that despite Theorem 2.10 it is not permissible to replace $U^{g_1} U U^{g_1}$ by U^{g_2} as the first term in the uppermost submatrix of A^{g_2}, since U^{g_1} must be the same matrix throughout, i.e. any U^{g_2} will not do. However, if we take $U^{g_1} = U^{g_2}$, which will give a more natural-looking result, then it is in order to have U^{g_2} as the first term.

Incidentally, the expressions for B^{g_1} and B^{g_2} given in the above theorems illustrate that in certain restricted cases it is possible to obtain expressions for g-inverses of a four-way partitioned matrix.

THEOREM 3.3 Let A be partitioned as in [3.2]; then

$$A^{g_3} = \begin{bmatrix} U^{g_3} - U^{g_3} V C^{g_3} \\ C^{g_3} \end{bmatrix},$$

where now $C = (I - UU^{g_3})V$.

Proof: Postmultiplication of A by P, where

$$P = \begin{bmatrix} I & -U^{g_3}V \\ 0 & I \end{bmatrix},$$

gives (defining B) $B = [U \quad C]$. Thus

$$B'B = \begin{bmatrix} U'U & 0 \\ 0 & C'C \end{bmatrix},$$

since $U'C = U'V - U'\underline{UU}^{g_3}V = U'V - U'(U^{g_3})'U'V = 0$, and by [3.1]

$$(B'B)^{g_1} = \begin{bmatrix} (U'U)^{g_1} & 0 \\ 0 & (C'C)^{g_1} \end{bmatrix}.$$

Now by [2.10] $B^{g_3} \equiv (B'B)^{g_1}B'$, so that

$$B^{g_3} = \begin{bmatrix} U^{g_3} \\ C^{g_3} \end{bmatrix},$$

and by Theorem 2.16, Corollary 2,

$$A^{g_3} = PB^{g_3} = \begin{bmatrix} U^{g_3} - U^{g_3}VC^{g_3} \\ C^{g_3} \end{bmatrix}.$$

This result is also proved by Chipman (1968).

Corollary If the columns of U form a basis for the column-space of A, so that $C = 0$, then

$$A^{g_3} = \begin{bmatrix} (U'U)^{-1}U' \\ 0 \end{bmatrix} = \begin{bmatrix} U^g \\ 0 \end{bmatrix}.$$

From Theorem 3.3 it is now apparent that simpler g_1- and g_2-inverses of $[U \quad V]$ than those derived in Theorems 3.1 and 3.2 are obtainable by replacing U^{g_3} by U^{g_1} and U^{g_2} respectively.

In the following theorem an expression for $[U \quad V]^g$ is derived. Since the method of Theorems 3.1–3.3 is no longer available for the g-inverse in view of the restricted nature of Theorem 2.16, Corollary 4, the theorem is proved from first principles.

THEOREM 3.4 Consider A to be partitioned as [3.2]. Then

$$A^g = \begin{bmatrix} U^g - U^g V(C^g + K) \\ C^g + K \end{bmatrix}, \qquad [3.4]$$

where

$$C = (I - UU^g)V, \qquad [3.5]$$

and

$$K = (I - C^g C)[I + (I - C^g C)V'(U^g)'U^g V(I - C^g C)]^{-1} V'(U^g)'U^g(I - VC^g).$$

Proof: Let A^g be conformably partitioned as

$$A^g = \begin{bmatrix} X \\ Y \end{bmatrix};$$

then

$$AA^g = UX + VY, \qquad [3.6]$$

and

$$A^g A = \begin{bmatrix} XU & XV \\ YU & YV \end{bmatrix}. \qquad [3.7]$$

Since $AA^g A = A$, the following relations hold:

$$AA^g U = U, \qquad [3.8]$$

$$AA^g V = V. \qquad [3.9]$$

Transposing [3.8] and multiplying on the left by $(U'U)^g$ gives, by Theorem 1.2(i),

$$U^g AA^g = U^g.$$

Furthermore, by Theorem 1.2(e), $A^g = A''(A^g)'A^g$, so that

$$X = U'X'X + U'Y'Y,$$

whence

$$\begin{aligned} U^g UX &= U^g UU'X'X + U^g UU'Y'Y \\ &= U'X'X + U'Y'Y \\ &= X. \end{aligned}$$

It follows that premultiplication of [3.6] by U^g gives $U^g = X + U^gVY$. Thus

$$A^g = \begin{bmatrix} U^g - U^gVY \\ Y \end{bmatrix}, \qquad [3.10]$$

and it remains to determine Y.

Now from [3.5] and [3.10],

$$AA^g = UU^g + (I - UU^g)VY = UU^g + CY. \qquad [3.11]$$

Premultiplication of C by U^g gives $U^gC = 0$, and thus, by Theorem 2.21,

$$C^gU = 0. \qquad [3.12]$$

Moreover, by [3.8] and [3.9], $AA^gC = AA^gV - AA^gUU^gV = V - UU^gV = C$, or $C' = C'AA^g$, whence $C^gAA^g = C^g$. Premultiplication of [3.11] by C^g thus gives $C^gCY = C^g$, so that

$$CY = CC^g. \qquad [3.13]$$

Also,

$$CC^gC = CC^g(V - UU^gV) = CC^gV = C, \text{ whence}$$

$$C^gV = C^gC. \qquad [3.14]$$

Thus $CYV = CC^gV = C$, and since YV is symmetric, $YVC' = C'$, or

$$YVC^g = C^g. \qquad [3.15]$$

It is readily verified from the four defining relations for A^g in the form [3.10] that Y must satisfy

$$YUU^g + YCY = Y,$$

or, by [3.13],

$$YUU^g + YCC^g = Y, \qquad [3.16]$$

$$U^gVYU = (U^gVYU)', \qquad [3.17]$$

and

$$(YU)' = U^gV(I - YV). \qquad [3.18]$$

These three conditions upon Y will be used to show that $Y = C^g + K$ yields the required form of A^g. Now, from [3.5] and [3.18]

$$\begin{aligned}(YU)' &= U^gV[I - Y(UU^gV + C)] \\ &= U^gV - U^gVYUU^gV - U^gVYC,\end{aligned}$$

and, since by [3.12] and [3.13]

$$(I - C^g C)YU = YU - C^g \underline{CYU} = YU - C^g CC^g YU = Y,$$
$$(YU)' = (U^g V - \underline{U^g VYUU^g} V)(I - C^g C).$$

Transposing, we have from [3.17],

$$YU = (I - C^g C)V'(U^g)' - (I - C^g C)V'(U^g)'U^g VYU,$$

and a further substitution of $(I - C^g C)YU$ for YU on the right-hand side yields

$$[I + (I - C^g C)V'(U^g)'U^g V(I - C^g C)]YU = (I - C^g C)V'(U^g)'.$$

Multiplying on the right by $U^g(I - VC^g)$ now gives, by [3.14], [3.15], and [3.16],

$$N(Y - C^g) = (I - C^g C)V'(U^g)'U^g(I - VC^g),$$

where N is the positive definite matrix

$$I + (I - C^g C)V'(U^g)'U^g V(I - C^g C).$$

Thus

$$Y = C^g + N^{-1}(I - C^g C)V'(U^g)'U^g(I - VC^g).$$

The expression for A^g given in [3.4] follows upon noting that $(I - C^g C)$ commutes with N, and hence with N^{-1}.

The result of this theorem was stated by Cline (1964a). It is clear that one might also have expressed A^g in terms of the matrix $C^* = (I - VV^g)U$ and the corresponding matrices N^* and K^*.

Although the expression [3.4] is excessively cumbersome and unlikely to be of much theoretical or practical use, it possesses several special cases which are of interest. These are presented in the following four corollaries, each of which was given by Cline, although his conditions are differently expressed.

Corollary 1 If and only if $U^g VC^g = 0$,

$$A^g = \begin{bmatrix} U^g - U^g VN_1^{-1}V'(U^g)'U^g \\ C^g + N_1^{-1}V'(U^g)'U^g \end{bmatrix}, \qquad [3.19]$$

where $N_1 = I + V'(U^g)'U^g V$. (Cline expressed this condition in the equivalent form $C^g CV'(U^g)'U^g V = 0$.)

Proof: That [3.4] reduces to [3.19] when $U^g VC^g = 0$ is obvious. Conversely, if [3.19] holds, then $AA^g A = A$ implies that

$$UU^g V - UU^g VN_1^{-1}V'(U^g)'U^g V + VC^g V + VN_1^{-1}V'(U^g)'U^g V = V. \qquad [3.20]$$

Premultiplication of [3.20] by U^g gives $U^g VC^g V = 0$, whence, by [3.14], $U^g VC^g C = 0$ or $U^g VC^g = 0$.

Corollary 2 If and only if V is contained in the column-space of U,

$$A^g = \begin{bmatrix} U^g - U^g V N_1^{-1} V'(U^g)' U^g \\ N_1^{-1} V'(U^g)' U^g \end{bmatrix}. \qquad [3.21]$$

Proof: By Theorem 1.5, $UU^g V = V$, so that $C = 0$ and hence $C^g = 0$. Corollary 1 therefore applies and [3.19] reduces further to [3.21]. If [3.21] holds, we arrive at a condition which is the same as [3.20] except that the term $VC^g V$ does not appear. Since $I - N_1^{-1} = N_1^{-1} V'(U^g)' U^g V$, this condition reduces to $CN_1^{-1} = 0$, i.e. $C = 0$, or $V = UU^g V$.

Corollary 3 If and only if $VC^g V = V$,

$$A^g = \begin{bmatrix} U^g - U^g VC^g \\ C^g \end{bmatrix}. \qquad [3.22]$$

Proof: If $VC^g V = V$, then by [3.14] $VC^g C = V$, so that $K = 0$ and A^g is as in [3.22]. If [3.22] holds, then $(A^g A)' = A^g A$ implies $U^g V(I - C^g C) = 0$, or $U^g V = U^g VC^g C$. Thus

$$\begin{aligned} C = V - UU^g V &= V - UU^g VC^g V \\ &= V - (V - C)C^g C \\ &= V - VC^g C + C, \end{aligned}$$

whence $VC^g C - V = 0$ and so $V = VC^g C = VC^g V$.

Attention is drawn to the similarity of [3.22] to the g_3-inverse of Theorem 3.3.

Corollary 4 If and only if $C = V$,

$$A^g = \begin{bmatrix} U^g \\ V^g \end{bmatrix}. \qquad [3.23]$$

Proof: If $C = V$, then $VC^g V = V$ and Corollary 3 is applicable. However, $U^g VC^g = U^g CC^g = 0$, and [3.22] reduces to [3.23]. Conversely, if [3.23] holds, then $AA^g A = A$ implies $UU^g V + VV^g V = V$. Hence $UU^g V = 0$ and $C = V$.

When the columns of U form a basis for the column-space of A

(which implies, of course, that $C = 0$), it is possible to obtain alternative representations for A^g. Several of these expressions are, however, computational formulae which have been suggested in the literature, and for this reason this particular case will be dealt with in Chapter 4.

3.3 Partitioned positive semidefinite matrices

Rohde (1964, pp. 47–51) has developed expressions for the four types of g-inverses of a positive semidefinite matrix $S = A'A$ partitioned as

$$S = \begin{bmatrix} S_{11} & S_{12} \\ S_{21} & S_{22} \end{bmatrix} = \begin{bmatrix} A_1'A_1 & A_1'A_2 \\ A_2'A_1 & A_2'A_2 \end{bmatrix}.$$

In this section Rohde's results will be described and, in the case of higher order g-inverses, slight extensions will be made.

The approach used by Rohde is equivalent to the consideration of the generalized equivalent reduction $P_1SP_2 = B$, where

$$B = P_1SP_2 = \begin{bmatrix} I & -S_{12}S_{22}^{g_1} \\ 0 & I \end{bmatrix} S \begin{bmatrix} I & 0 \\ -S_{22}^{g_1}S_{21} & I \end{bmatrix}$$

$$= \begin{bmatrix} Q & 0 \\ 0 & S_{22} \end{bmatrix}, \qquad [3.24]$$

where $Q = S_{11} - S_{12}S_{22}^{g_1}S_{21}$. The reduction is achieved by virtue of the relations

$$S_{11}S_{11}^{g_1}S_{12} = S_{12} \quad \text{and} \quad S_{21}S_{11}^{g_1}S_{11} = S_{21}, \qquad [3.25]$$

which follow by [2.5] and [2.6]. It is worth noting that, by [2.7], Q may be written as $Q = A_1'(I - A_2A_2^g)A_1$, and is therefore symmetrical and independent of the choice of g-inverse of S_{22}. Since $(S_{22}^{g_1})'$ is also a g_1-inverse of S_{22}, it follows that the same reduction occurs if P_2 is replaced by P_1', so that the transformation is a generalized congruent reduction of the type PSP' rather than P_1SP_2. This will also be so if a symmetric g_1-inverse of S_{22} is chosen.

By Theorem 1.3, $S^{g_1} = P_2B^{g_1}P_1$, and the choice of B^{g_1} of the type [3.1], namely

$$\begin{bmatrix} Q^{g_1} & 0 \\ 0 & S_{22}^{g_1} \end{bmatrix},$$

yields

$$S^{g_1} = \begin{bmatrix} Q^{g_1} & -Q^{g_1} S_{12} S_{22}^{g_1} \\ -S_{22}^{g_1} S_{21} Q^{g_1} & S_{22}^{g_1} + S_{22}^{g_1} S_{21} Q^{g_1} S_{12} S_{22}^{g_1} \end{bmatrix}. \quad [3.26]$$

When S is nonsingular, S_{22} is nonsingular and [3.26] reduces to a commonly given expression for the inverse of a partitioned matrix, namely Schur's Identity.

By Theorem 1.8 it follows that the choice of B^{g_2} of the type

$$\begin{bmatrix} Q^{g_2} & 0 \\ 0 & S_{22}^{g_2} \end{bmatrix}$$

gives

$$S^{g_2} = P_2 B^{g_2} P_1 = \begin{bmatrix} Q^{g_2} & -Q^{g_2} S_{12} S_{22}^{g_1} \\ -S_{22}^{g_1} S_{21} Q^{g_2} & S_{22}^{g_2} + S_{22}^{g_1} S_{21} Q^{g_2} S_{12} S_{22}^{g_1} \end{bmatrix}. \quad [3.27]$$

As regards the higher order g-inverses of S, Rohde considered the following products, using [2.5] and [2.6] to obtain relatively simplified forms:

$$SS^{g_1} = \begin{bmatrix} QQ^{g_1} & (I - QQ^{g_1})S_{12} S_{22}^{g_1} \\ 0 & S_{22} S_{22}^{g_1} \end{bmatrix} \quad [3.28]$$

$$S^{g_1} S = \begin{bmatrix} Q^{g_1} Q & 0 \\ S_{22}^{g_1} S_{21} (I - Q^{g_1} Q) & S_{22}^{g_1} S_{22} \end{bmatrix}. \quad [3.29]$$

To obtain a g_3-inverse of S, it is required that [3.28] be symmetric. The diagonal submatrices will certainly be symmetric if B^{g_1} is taken as

$$\begin{bmatrix} Q^{g_3} & 0 \\ 0 & S_{22}^{g_3} \end{bmatrix},$$

and clearly

$$P_2\begin{bmatrix} Q^{g_3} & 0 \\ 0 & S_{22}^{g_3} \end{bmatrix} P_1 = S^{g_2}.$$

It remains to establish conditions under which $(I - QQ^{g_1})S_{12}S_{22}^{g_1} = 0$. Rohde shows that if S_{11} is nonsingular and

$$r(S_{11}) + r(S_{22}) = r(S), \quad [3.30]$$

then Q is nonsingular and the desired result holds. However, it will now be shown that [3.30] alone is both necessary and sufficient for $(I - QQ^{g_1})S_{12}S_{22}^{g_1} = 0$.

Since $Q = A_1'(I - A_2A_2^{g})A_1$, the column-space of Q is contained in that of A_1'. If $r(Q) = r(A_1')$ the two column-spaces are equal and A_1' may be written in the form $A_1' = QE$ for some matrix E. Thus $QQ^{g_1}S_{12} = QQ^{g_1}A_1'A_2 = QQ^{g_1}QEA_2 = S_{12}$, i.e. the condition $r(A_1') = r(Q)$ is sufficient for $(I - QQ^{g_1})S_{12}S_{22}^{g_1} = 0$. Conversely, if $(I - QQ^{g_1})S_{12}S_{22}^{g_1} = 0$, postmultiplication by S_{21} gives $(I - QQ^{g_1})(S_{11} - Q) = 0$, or $QQ^{g_1}S_{11} = S_{11}$, which implies that A_1' is contained in the column-space of Q, and hence that $r(A_1') = r(Q)$. Thus the condition $r(A_1') = r(Q)$ is necessary and sufficient. That [3.30] is equivalent follows from the fact that $r(S) = r(B) = r(Q) + r(S_{22})$.

Hence, if [3.30] holds, then

$$S^{g_3} = P_2B^{g_3}P_1 = \begin{bmatrix} Q^{g_3} & -Q^{g_3}S_{12}S_{22}^{g_1} \\ -S_{22}^{g_1}S_{21}Q^{g_3} & S_{22}^{g_3} + S_{22}^{g_1}S_{21}Q^{g_3}S_{12}S_{22}^{g_1} \end{bmatrix}. \quad [3.31]$$

It would appear that this result has also been obtained by Zelen & Federer (1965).

In similar fashion, consideration of [3.29] shows that it is possible to construct the g-inverse of a partitioned matrix S in the form

$$S^{g} = P_2B^{g}P_1 = \begin{bmatrix} Q^{g} & -Q^{g}S_{12}S_{22}^{g_1} \\ -S_{22}^{g_1}S_{21}Q^{g} & S_{22}^{g} + S_{22}^{g_1}S_{21}Q^{g}S_{12}S_{22}^{g_1} \end{bmatrix}, \quad [3.32]$$

provided, of course, that [3.30] holds.

If any of the expressions [3.27], [3.31], and [3.32] were used as a computational method it would be completely unnecessary to compute both $S_{22}^{g_1}$ and, for example, $S_{22}^{g_2}$, since by [1.24] every g_2-inverse of S_{22} is also a g_1-inverse of S_{22}. Now the computation of [3.26] consists mainly of finding a g_1-inverse of S_{22}. In some situations it may, however, be simpler to obtain $S_{11}^{g_1}$, in which case the choice of P_1 as

$$P_1 = \begin{bmatrix} I & 0 \\ -S_{21}S_{11}^{g_1} & I \end{bmatrix},$$

with a corresponding form for P_2, leads to

$$S^{g_1} = \begin{bmatrix} S_{11}^{g_1} + S_{11}^{g_1}S_{12}Z^{g_1}S_{21}S_{11}^{g_1} & -S_{11}^{g_1}S_{12}Z^{g_1} \\ -Z^{g_1}S_{21}S_{11}^{g_1} & Z^{g_1} \end{bmatrix}, \qquad [3.33]$$

where $Z = S_{22} - S_{21}S_{11}^{g_1}S_{12}$. In a similar manner it is possible to obtain partitioned forms for S^{g_2}, and subject to [3.30] for S^{g_3} and S^{g} in terms of g-inverses of S_{11} and Z.

3.4 Bordered matrices

Bordered matrices of the form

$$M = \begin{bmatrix} S & L' \\ L & 0 \end{bmatrix}, \qquad [3.34]$$

where $S = A'A$, frequently arise in statistical theory, in particular in the theory of constrained linear models. In this section expressions for g-inverses of the matrix M will be derived.

Once again, the expressions are obtained by seeking nonsingular matrices P_1 and P_2 such that $P_1MP_2 = B$, where B is a quasi-diagonal matrix. The result [3.1] is then used to obtain B^{g_1}, and $M^{g_1} = P_2B^{g_1}P_1$.

The row-operations on M may be considered to be performed in two stages. Firstly, M may be premultiplied by F, where

$$F = \begin{bmatrix} I & L' \\ 0 & I \end{bmatrix},$$

to give

$$FM = \begin{bmatrix} S + L'L & L' \\ L & 0 \end{bmatrix}.$$

Now let

$$S + L'L = K, \qquad [3.35]$$

and

$$LK^{g_1}L' = R, \qquad [3.36]$$

and consider the reduction $E(FM)P_2$, where

$$E = \begin{bmatrix} I & 0 \\ -LK^{g_1} & I \end{bmatrix} \quad \text{and} \quad P_2 = \begin{bmatrix} I & -K^{g_1}L' \\ 0 & I \end{bmatrix}.$$

The product $EF = P_1$ is, of course, nonsingular. Furthermore, the row-spaces of A and L are each contained in the row-space of K. Thus, by Theorem 1.5 and its corollary, the following relations are apparent:

$$LK^{g_1}K = L \qquad [3.37]$$

$$KK^{g_1}L' = L' \qquad [3.38]$$

$$AK^{g_1}K = A \qquad [3.39]$$

$$KK^{g_1}A' = A'. \qquad [3.40]$$

It now follows from [3.36], [3.37], and [3.38] that

$$P_1MP_2 = \begin{bmatrix} K & 0 \\ 0 & -R \end{bmatrix},$$

and by [3.1] that

$$M^{g_1} = P_2\begin{bmatrix} K^{g_1} & 0 \\ 0 & -R^{g_1} \end{bmatrix}P_1, \qquad [3.41]$$

with

$$P_1 = \begin{bmatrix} I & L' \\ -LK^{g_1} & I-R \end{bmatrix}.$$

In reducing [3.41] it is noted that by [3.37] and Theorem 2.17, $R = LK^{g_1}L'$ is unique, positive semidefinite and has the same rank as L, whence by Theorem 2.17, Corollary 1,

$$L'R^{g_1}R = L'. \qquad [3.42]$$

On completion of the reduction the following result has been proved :-

THEOREM 3.5 Let S be any $k \times k$ positive semidefinite matrix and L any $q \times k$ matrix ; then

$$\begin{bmatrix} S & L' \\ L & 0 \end{bmatrix}^{g_1} = \begin{bmatrix} K^{g_1} - K^{g_1}L'R^{g_1}LK^{g_1} & K^{g_1}L'R^{g_1} \\ R^{g_1}LK^{g_1} & R^{g_1}R - R^{g_1} \end{bmatrix}, \qquad [3.43]$$

where K and R are as in [3.35] and [3.36] respectively.

It may be noted that [3.43] with the suffix 1 changed to 2 or 3 throughout, or deleted, provides expressions for M^{g_2}, M^{g_3}, and M^{g} also. The g_2-case follows similarly to [3.41], since

$$M^{g_2} = P_2 \begin{bmatrix} K^{g_2} & 0 \\ 0 & -R^{g_2} \end{bmatrix} P_1,$$

by Theorem 1.8. This theorem does not hold, however, in general for a g_3-inverse or the g-inverse, but the symmetry of MM^{g_3} and M^gM with

$$M^{g_3} = P_2 \begin{bmatrix} K^{g_3} & 0 \\ 0 & -R^{g_3} \end{bmatrix} P_1 \quad \text{and} \quad M^{g} = P_2 \begin{bmatrix} K^{g} & 0 \\ 0 & -R^{g} \end{bmatrix} P_1$$

may nevertheless be verified. Of course, when obtaining M^{g_2}, M^{g_3}, and M^g, K^{g_1} in P_1 and P_2 must change accordingly, but since by Theorem 2.17, Corollary 2, $L'R^{g_1}L$ is unique, there is no need to change R^{g_1} in the leading submatrix of [3.43]. The expression for M^{g_3} now obtained is simpler than that derived by Rohde (1964, p. 71). This extension of Theorem 3.5 also applies to Corollaries 1, 2, and 3.

If certain relations between L and S exist, then [3.43] may be considerably simplified.

Corollary 1 If L is contained in the row-space of S,

$$M^{g_1} = \begin{bmatrix} S^{g_1} - S^{g_1}L'(LS^{g_1}L')^{g_1}LS^{g_1} & S^{g_1}L'(LS^{g_1}L')^{g_1} \\ (LS^{g_1}L')^{g_1}LS^{g_1} & -(LS^{g_1}L')^{g_1} \end{bmatrix}. \qquad [3.44]$$

Proof: By Theorem 2.18,

$$\begin{aligned} R &= L(S + L'L)^{g_1}L' \\ &= LS^{g_1}L' - LS^{g_1}L'(I + LS^{g_1}L')^{-1}LS^{g_1}L' \\ &= LS^{g_1}L'[I - (I + LS^{g_1}L')^{-1}LS^{g_1}L'] \\ &= LS^{g_1}L'(I + LS^{g_1}L')^{-1} \\ &= (I + LS^{g_1}L')^{-1}LS^{g_1}L', \end{aligned}$$

whence

$$\left.\begin{aligned} R^{g_1} &= (LS^{g_1}L')^{g_1}(I + LS^{g_1}L') \\ &= (I + LS^{g_1}L')(LS^{g_1}L')^{g_1}. \end{aligned}\right\} \qquad [3.45]$$

Clearly the submatrix $R^{g_1}R - R^{g_1}$ in [3.43] reduces to $-(LS^{g_1}L')^{g_1}$. By using Theorem 2.18 and [3.45] it can be shown in similar fashion that the other submatrices in [3.43] reduce to the appropriate submatrices in [3.44].

The expression [3.44] can also be derived by noting that, for

$$P_1 = \begin{bmatrix} I & 0 \\ -LS^{g_1} & I \end{bmatrix} \quad \text{and} \quad P_2 = \begin{bmatrix} I & -S^{g_1}L' \\ 0 & I \end{bmatrix},$$

$$P_1MP_2 = \begin{bmatrix} S & 0 \\ 0 & -LS^{g_1}L' \end{bmatrix},$$

whence

$$M^{g_1} = P_2 \begin{bmatrix} S^{g_1} & 0 \\ 0 & -(LS^{g_1}L')^{g_1} \end{bmatrix} P_1,$$

which is the same as [3.44].

Corollary 2 If $r(L) = q \leqslant k$, and L is contained in the row-space of S,

$$M^{g_1} = \begin{bmatrix} S^{g_1} - S^{g_1}L'(LS^{g_1}L')^{-1}LS^{g_1} & S^{g_1}L'(LS^{g_1}L')^{-1} \\ (LS^{g_1}L')^{-1}LS^{g_1} & -(LS^{g_1}L')^{-1} \end{bmatrix}. \qquad [3.46]$$

Proof: Under the given conditions, $LS^{g_1}L'$ is nonsingular (Theorem 2.17) and the result follows from the first corollary.

If S is nonsingular, then [3.46] reduces to the expression for the regular inverse of M as presented, for example, by Plackett (1960, p. 67).

Corollary 3 If L is a $q \times k$ matrix complementary to S, then

$$M^{g} = \begin{bmatrix} K^{-1}SK^{-1} & K^{-1}L' \\ LK^{-1} & 0 \end{bmatrix}. \qquad [3.47]$$

Proof: As shown in Theorem 2.19(a), K is nonsingular. Also $R = LK^{-1}L'$ is idempotent since $LK^{-1}L'LK^{-1}L' = LK^{-1}(K-S)K^{-1}L' = LK^{-1}L'$ by Theorem 2.19(a). Now $L'R^{g_1}L$ is unique (Theorem 2.17, Corollary 2), i.e. it may be written as $L'R^{g}L$, and by Example 1.1 $R^{g} = R$. The leading submatrix in [3.43] therefore becomes $K^{-1} - K^{-1}L'LK^{-1}L'LK^{-1}$, or $K^{-1} - K^{-1}L'LK^{-1}$, since $K^{-1}L' = L^{g_3}$. The substitution $L'LK^{-1} = I - SK^{-1}$ gives the leading submatrix in [3.47]. It is convenient to choose $R^{g_1} = R^{g} = R$, so that $R^{g_1}R - R^{g_1} = R^2 - R = 0$, and $K^{g_1}L'R^{g_1} = K^{-1}L'LK^{-1}L' = K^{-1}L'$ by Theorem 2.19(c).

Corollary 4 If L is as in Corollary 3, but with $\mathrm{r}(L) = q \leqslant k$, it is well known that M has full rank, i.e.

$$M^{-1} = \begin{bmatrix} K^{-1}SK^{-1} & K^{-1}L' \\ LK^{-1} & 0 \end{bmatrix}. \qquad [3.48]$$

A direct derivation of [3.48], using the results of Theorem 2.19, has been given by Rayner & Pringle (1967). Goldman & Zelen (1964) and Rohde (1964) have derived alternative expressions for the submatrices of M^{-1} in terms of the matrices S, L, and B, where S is as above but L and B are matrices such that $SB = 0$ and $\det(LB) \neq 0$. However, it is evident from Theorem 2.19(a) that B may be taken as $K^{-1}L' = L^{g_3}$, so that $LL^{g_3} = I$, with considerable simplification of the algebra.

Goldman & Zelen established that the leading submatrix in M^{-1} is S^{g_2}. This is also evident from Theorem 2.10, as $K^{-1} = S^{g_1}$.

Corollary 5 If S is as in Corollary 4 and $\mathrm{r}(L) = q \leqslant k$, but L is now an orthogonal complement of A, i.e. $AL' = 0$; then

$$M^{-1} = \begin{bmatrix} S^{g} & L^{g} \\ (L^{g})' & 0 \end{bmatrix}.$$

This result, which is implicit in the work of Goldman & Zelen, follows directly from Corollary 4 and Theorem 2.20.

In a recent paper Khatri (1968) has studied the properties of g_1-inverses of the matrix M, for the cases L any matrix and L complementary to S. However, he does not derive any explicit expressions for M^{g_1}.

3.5 Some remarks on generalized inverses of partitioned and bordered matrices

In both of the previous two sections the expressions for a g_1-inverse of the appropriate matrix A (say) were derived by an equivalent reduction of A to quasi-diagonal form, i.e. $P_1AP_2 = B$. The relation [3.1] was then used to obtain a g_1-inverse of B and hence of A.

The generalized equivalent reduction [3.24] is subject only to the conditions [3.25] and therefore applies to any matrix A partitioned as in [1.18] and such that A_{12} and A_{21} are contained in the column-space and row-space, respectively, of A_{11} (Theorem 1.5 and Corollary). Hence the g_1-inverse of such a matrix may also be obtained as $P_2B^{g_1}P_1$, where B is quasi-diagonal. The expressions for g-inverses of bordered and partitioned positive semidefinite matrices are therefore special cases of this more general result, but they are the ones which occur naturally in statistical theory. Actually Corollary 1 of Theorem 3.5 follows immediately from [3.33] with suitable notational changes, since L is contained in the row-space of S. The higher order g-inverses are also similar to [3.31] and [3.32] because [3.30] holds.

It is worthwhile mentioning that the relation [3.1] does not provide a complete characterization of g_1-inverses of quasi-diagonal matrices, and the expressions for g_1-inverses of bordered and partitioned matrices so derived are therefore only a subset of the set of all g_1-inverses. The construction of general forms for g_1-inverses may be achieved as follows:

Let

$$P_1AP_2 = B = \begin{bmatrix} B_1 & 0 \\ 0 & B_2 \end{bmatrix}$$

and let

$$B^{g_1} \equiv \begin{bmatrix} C_1 & C_2 \\ C_3 & C_4 \end{bmatrix}.$$

Then the submatrices C_i must satisfy

$$B_1C_1B_1 = B_1,$$

$$B_2C_4B_2 = B_2,$$

$$B_1C_2B_2 = 0, \qquad [3.49]$$

$$B_2C_3B_1 = 0. \qquad [3.50]$$

The first two relations imply, of course, that $C_1 = B_1^{g_1}$ and $C_4 = B_2^{g_1}$, while C_2 and C_3 are solutions to [3.49] and [3.50] respectively. The general solutions to [3.49] and [3.50] are, by Theorem 1.6,

$$C_2 = U - B_1^{g_1}B_1UB_2B_2^{g_1}, \qquad [3.51]$$

and

$$C_3 = V - B_2^{g_1}B_2VB_1B_1^{g_1}, \qquad [3.52]$$

where U and V are arbitrary.

We can therefore construct, from the above results, general expressions for g_1-inverses of the matrices $S = \begin{bmatrix} S_{11} & S_{12} \\ S_{21} & S_{22} \end{bmatrix}$ and $M = \begin{bmatrix} S & L' \\ L & 0 \end{bmatrix}$. For example, from [3.51], [3.52], and [3.24], we may write

$$\begin{bmatrix} S_{11} & S_{12} \\ S_{21} & S_{22} \end{bmatrix}^{g_1} \equiv P_2 \begin{bmatrix} Q^{g_1} & U - Q^{g_1}QUS_{22}S_{22}^{g_1} \\ V - S_{22}^{g_1}S_{22}VQQ^{g_1} & S_{22}^{g_1} \end{bmatrix} P_1,$$

where U and V are arbitrary, and P_1 and P_2 are as in [3.24].

In statistical applications it is more convenient to approach the problems of non-uniqueness of g_1-inverses and the generality of derived statistical results by using the theory of the general solution to linear equations (§1.4). Thus the general solutions [1.16] and [1.17] may be constructed from any convenient particular g_1-inverse of A.

Chapter 4

METHODS OF COMPUTING GENERALIZED INVERSES

4.1 Introduction

In the computation of g-inverses a range of situations may be recognized: (a) The matrix is a very small one with simple figures suitable for mental calculations, as in little textbook examples. (b) The matrix is small without special features and may be "inverted" on a desk calculating machine. (c) The matrix has special features which are readily apparent, as in some statistical situations. Partitioned matrix formulae may then be utilized opportunistically to reduce the calculations to the mental or desk-calculator level, or to permit them to be handled on a computer of limited capacity. (d) The matrix is large, with or without special features, and the "inversion" is done by computer.

Although it would seem to be the last category which should deserve the greatest attention in the literature, it is precisely here that a relative dearth of papers exists. Most of the writing on the computational aspect of g-inverses is actually of a theoretical nature, i.e. it deals with alternative algebraic structures for the various types of g-inverses. This is probably because the usefulness of g-inverses, in statistical areas at least, is theoretical rather than practical. However, the appearance of one or two recent papers on purely numerical aspects may reflect a growing interest in this side of the subject.

In a work of this sort it is impossible to consider such details as computer programs or the relative efficiencies of various methods, and the present contribution will therefore tend to follow the established pattern.

Nevertheless it is important to discuss one general point which may not affect little textbook examples, but which must be borne in mind in practical calculations. The formulae presented are mostly such that the first r rows or columns of the matrix (where r is its rank) form a basis, or there is a leading $r \times r$ submatrix of full rank. This may be through computational necessity or notational convenience. In the former case it is necessary to confirm that the matrix in question is in the appropriate form, or its rows and columns must be re-ordered so that it is so. In either case recognition of the rank of the matrix, or when a row or column is dependent on its predecessors, cannot be avoided. In the latter case the determination of rank is implicit in the calculations, and dependent rows or columns are passed over as they arise without any actual re-ordering. This also applies to those formulae in which no special order of rows and columns appears to be necessary, either computationally or notationally. Hence, save in exceptional cases (e.g. the g-inverse of a matrix of full row- or column-rank, or the use of Theorem 2.19 (c) with H known), the computations require recognition of dependent rows and columns, and there is therefore the problem of accurate detection, in the presence of rounding errors, of zero pivotal elements or a zero row or column.

The problem is particularly pertinent to the floating-point arithmetic of the computer, and Golub & Kahan (1964), in reviewing this situation, point out that it can arise even when the matrix appears to be well-conditioned. In their opinion a satisfactory solution has yet to be obtained, although they express some faith in their method for the g-inverse based on the decomposition of Theorem 1.1 (see next section), where the difficulty manifests itself in the recognition of a zero latent root. Tewarson (1967–8) also discusses the problem, referring in particular to a technique due to Osborne (1965) for the recognition of zero rows. Healy (1968a) is another author to highlight the problem of nearly-zero rows.

Golub & Kahan suggest the advisability, as a palliative measure, of scaling the matrix in some manner so as to allow each row and column "to communicate its proper significance to the calculation". For example, they recommend that the matrix be scaled so that its rows have approximately the same norm, and similarly for the columns.

In some statistical applications there may be no problem in determining the rank of the matrix. For example, in the least-squares analysis of certain experimental design models, the redundant rows and columns are evident from the nature of the model, but then in all likelihood the computing problem would revert to category (c) mentioned earlier.

4.2 The g-inverse

Although many different formulae for computing the g-inverse have been published in the literature, most of them rest on the following basic method. Briefly, an arbitrary $n \times k$ matrix A, of rank r, is partitioned as $A = [U \quad V]$, where U is $n \times r$ of rank r. If the first r columns of A are not linearly independent, then there exists a permutation matrix P such that $A^* = AP = [U \quad V]$, where U has the desired properties. Since $(A^*)^g = P'A^g$, there is very little loss in generality in assuming that the first r columns of A are linearly independent. Authors such as Boot (1963), Chipman (1964), and Rust *et al.* (1966) then use the fact that U has full column-rank to express A^g in terms of ordinary inverses.

However, the above partitioning is merely a special case of the more general partitioning considered in § 3.2, and the various forms for A^g may be obtained directly from Theorem 3.4 and its corollaries.

Since U has full column-rank, $U^g U = I$ (cf. § 2.6) and, furthermore, V has the unique representation $V = UK$, where K is a matrix of dimension $r \times (k - r)$. Clearly $K = U^g V$. Now, since V is contained in the column-space of U, an expression for A^g may be obtained from [3.21]. Thus

$$A^g = \begin{bmatrix} \{I - K(I + K'K)^{-1}K'\}U^g \\ (I + K'K)^{-1}K'U^g \end{bmatrix}, \qquad [4.1]$$

where the substitution $U^g V = K$ has been made and $U^g = (U'U)^{-1}U'$. It is well known that $I - K(I + K'K)^{-1}K' = (I + KK')^{-1}$ and $(I + K'K)^{-1}K' = K'(I + KK')^{-1}$, so that [4.1] may also be expressed as

$$A^g = \begin{bmatrix} (I + KK')^{-1}U^g \\ K'(I + KK')^{-1}U^g \end{bmatrix}. \qquad [4.2]$$

The relative sizes of $k - r$ and r will determine whether [4.1] or [4.2] is the more convenient form. These expressions were derived by Rust *et al.* from the point of view of minimizing the length of the vector $\mathbf{b}$, subject to the restriction that $\mathbf{b}$ satisfy the equation $A\mathbf{b} = \mathbf{y}$. This result is discussed in Chapter 6.

These authors have also shown how the Schmidt orthogonalization process may be utilized to give U^g and K in a systematic fashion: By the Schmidt process, an $n \times r$ matrix U with independent columns may be factored as

$$U = ZT, \qquad [4.3]$$

where Z is an $n \times r$ column-orthonormal matrix and T is an $r \times r$ upper triangular matrix with positive diagonal elements. If $\mathbf{z}_i$ and $\mathbf{u}_i$ denote the r columns of Z and U respectively, then the $\mathbf{z}_i$ may be determined recursively as

$$\mathbf{z}_i = (\mathbf{c}_i'\mathbf{c}_i)^{-\frac{1}{2}}\mathbf{c}_i, \qquad [4.4]$$

where $$\mathbf{c}_i = \mathbf{u}_i - \sum_{j=1}^{i-1} (\mathbf{u}_i' \mathbf{z}_j)\mathbf{z}_j. \qquad [4.5]$$

When the process is applied to $A = [U \quad V]$, the matrix V becomes

$$\begin{aligned} V - ZZ'V &= V - UT^{-1}Z'V \\ &= V - UU^g V \quad \text{by Theorem 2.16, Corollary 3} \\ &= 0 \quad \text{by Theorem 1.5.} \end{aligned}$$

Thus the Schmidt orthogonalization process may be represented as postmultiplication by

$$\begin{bmatrix} T^{-1} & -U^g V \\ 0 & I_{k-r} \end{bmatrix} = \begin{bmatrix} T^{-1} & -K \\ 0 & I \end{bmatrix}.$$

Hence if A is augmented by r rows $[I_r \quad 0]$ and the same column operations are applied to the whole augmented matrix, we have

$$\begin{bmatrix} U & V \\ I_r & 0 \end{bmatrix} \begin{bmatrix} T^{-1} & -K \\ 0 & I \end{bmatrix} = \begin{bmatrix} Z & 0 \\ T^{-1} & -K \end{bmatrix}.$$

This gives K and $U^g = T^{-1}Z'$, leaving only the inversion of $I + KK'$ or $I + K'K$.

Thus expressions [4.1] and [4.2] effectively require the inversion of two matrices, as would also be the case if [1.8] were to be utilized for computational purposes. Boot (1963) has, however, developed an expression for A^g which requires the inversion of a single $r \times r$ matrix. Once again suppose that the columns of U form a basis for the column-space of A, and define the matrix Q by

$$\begin{aligned} Q &= [U'U \quad U'V] \qquad [4.6] \\ &= U'U[I \quad K], \end{aligned}$$

where $K = U^g V$ as before. Clearly Q has full row-rank and

$$(QQ')^{-1} = (U'U)^{-1}(I + KK')^{-1}(U'U)^{-1}$$

or

$$U'U(QQ')^{-1}U' = (I + KK')^{-1}U^g.$$

Thus [4.2] becomes

$$A^g = \begin{bmatrix} I \\ K' \end{bmatrix} U'U(QQ')^{-1}U' = Q'(QQ')^{-1}U'. \qquad [4.7]$$

Boot's proof of [4.7] was based on the result that A^g is that matrix G which minimizes $\mathrm{tr}(GG')$ subject to $A'AG = A'$ (cf. Chipman & Rao, 1964).

From [4.6] it can be seen that Q may also be expressed as $Q = U'A$, and thus [4.7] becomes $A^g = A'U(U'AA'U)^{-1}U'$, which is a computational form suggested by Chipman (1964).

Cline (1964a) has pointed out that Corollaries 2 and 3 of Theorem 3.4 are generalizations of a recursive algorithm for computing A^g developed by Greville (1960). To obtain his algorithm, Greville considered a matrix A, with k columns, to be partitioned as $A_k = [A_{k-1} \quad \mathbf{a}_k]$, where A_{k-1} denotes the submatrix consisting of the first $k - 1$ columns and $\mathbf{a}_k$ is a single column, and showed that

$$A_k^g = \begin{bmatrix} A_{k-1}^g - \mathbf{d}_k \mathbf{b}_k' \\ \mathbf{b}_k' \end{bmatrix},$$

where

$$\mathbf{d}_k = A_{k-1}^g \mathbf{a}_k,$$

$$\mathbf{b}_k = \begin{cases} \mathbf{c}_k^g & \text{if } \mathbf{c}_k \neq \mathbf{0} \\ (1 + \mathbf{d}_k'\mathbf{d}_k)^{-1}\mathbf{d}_k' A_{k-1}^g & \text{if } \mathbf{c}_k = \mathbf{0}, \end{cases}$$

and $\mathbf{c}_k = \mathbf{a}_k - A_{k-1}\mathbf{d}_k$. However, Greville's result follows directly from Theorem 3.4 since, for this particular partitioning of A, the matrix $C = (I - UU^g)V$ of [3.5] reduces to a column-vector, namely $\mathbf{c}_k$. If $\mathbf{c}_k \neq \mathbf{0}$, $C^gC = 1$, so that by [3.14] $VC^gV = V$ and Corollary 3 yields the desired expression for A^g. If $\mathbf{c}_k = \mathbf{0}$, $\mathbf{a}_k$ is in the column-space of A_{k-1} and Corollary 2 applies.

Penrose (1956) presented a method of computing A^g based on the result that, with a suitable arrangement of rows and columns, any matrix A of rank r can be expressed as

$$A = \begin{bmatrix} A_{11} & A_{12} \\ A_{21} & A_{21}A_{11}^{-1}A_{12} \end{bmatrix},$$

where A_{11} is $r \times r$ nonsingular. Let $\begin{bmatrix} I \\ A_{21}A_{11}^{-1} \end{bmatrix} = P_1$ and $[I \quad A_{11}^{-1}A_{12}] = P_2$,

whence $A = P_1 A_{11} P_2$. Since P_1 has full column-rank and P_2 full row-rank, $A^g = P_2^g A_{11}^{-1} P_1^g$ by Theorem 2.16, Corollary 5, with

$P_1^g = (P_1'P_1)^{-1}P_1'$ and $P_2^g = P_2'(P_2P_2')^{-1}$, i.e.

$$A^g = \begin{bmatrix} A_{11}' \\ A_{12}' \end{bmatrix} Q[A_{11}' \quad A_{21}'], \qquad [4.8]$$

where $Q = (A_{11}A_{11}' + A_{12}A_{12}')^{-1}A_{11}(A_{11}'A_{11} + A_{21}'A_{21})^{-1}$. Tewarson (1967–8) has shown that the quantities A_{11}^{-1}, $A_{11}^{-1}A_{12}$, and $A_{21}A_{11}^{-1}$ may be calculated by the well-known Gauss–Jordan elimination method.

Graybill *et al.* (1966) have suggested a method of computing A^g when A is symmetrical, which is based on the result $A^g = (A^{g_3})'AA^{g_3}$ (Theorem 2.14, Corollary). In their method A^g is calculated as $A^g = X'AX$, where X is any solution to $A^2X = A$. Now by [1.16] X must be of the form $(A^2)^{g_1}A + [I - (A^2)^{g_1}A^2]Z$. By [2.10] $(A^2)^{g_1}A = A^{g_3}$, and hence $X'AX = (A^{g_3})'AA^{g_3}$. They also suggest that when A is not symmetric and of order $n \times k$, A^g be calculated as $(A'A)^gA'$ (for $n \leqslant k$) or $A'(AA')^g$ (for $n \geqslant k$), where $(A'A)^g$ or $(AA')^g$ is obtained as above.

Actually $(A'A)^g$ can be computed (knowing A) as $A^g(A^g)'$ by using any of the methods discussed above for calculating the g-inverse of $A = [U \quad V]$. For example, from [4.2] the result

$$(A'A)^g = \begin{bmatrix} P & PK \\ K'P & K'PK \end{bmatrix}, \qquad [4.9]$$

where $P = (I + KK')^{-1}(U'U)^{-1}(I + KK')^{-1}$, is obtained. Since $K = U^gV$, P may be written as $P = U'UWU'U$, where

$$W = [(U'U)^2 + U'VV'U]^{-1}U'U[(U'U)^2 + U'VV'U]^{-1}.$$

Thus [4.9] becomes

$$(A'A)^g = \begin{bmatrix} U'UWU'U & U'UWU'V \\ V'UWU'U & V'UWU'V \end{bmatrix},$$

which is the same as [4.8] for this special case.

For a positive semidefinite matrix S (not given as $A'A$ with A known), a result due to Marsaglia (1964) and Rohde (1964, p. 57) may be used. These authors point out that S may be factored as $R'R$, where R has full row-rank, and that by [1.8], $S^g = R'(RR')^{-2}R$. Rohde (p. 69) has shown how the Doolittle technique may be used to compute $R'(RR')^{-2}R$ in a routine fashion similar to Aitken's triple product method (Aitken, 1937). A further discussion on Rohde's method may be found in Rohde & Harvey (1965). Rohde also shows how $(S^2)^{g_1}S$

can be similarly computed, i.e. S^{g_3}, and this could be used to obtain S^g as in the method of Graybill *et al.* These methods could be equally well described in terms of the Cholesky technique (§ 4.6) but in any case they seem unduly bulky.

Golub & Kahan (1964) have considered the computation of the g-inverse in the form given in Theorem 1.1, namely $A^g = Q'\Lambda^{-\frac{1}{2}}H_1$, where Q and H_1 are row-orthonormal matrices, and $\Lambda^{\frac{1}{2}}$ is a diagonal matrix with elements comprising the positive square roots of the non-zero latent roots of AA'. The problem here is that of accurately detecting the zero latent roots. However, Golub & Kahan feel that their method is not as seriously affected by the computational difficulties discussed earlier. In a later paper, Golub (1969) has discussed some general considerations concerning the numerical methods involved in various matrix decompositions (e.g. the Schmidt orthogonalization), iterative refinements, and ill-conditioned matrices. Although the results mostly concern nonsingular matrices, the paper has considerable bearing upon the methods considered in this chapter.

4.3 g_1-inverses

We consider here methods of computing matrices which satisfy only the condition $AGA = A$. A method, which will be loosely referred to as "sweep-out" or "pivotal condensation" appears to be the simplest, although not necessarily the most compact, way of finding A^{g_1}. This method, which has been used by Rao (1962, 1965) and Rohde (1964, pp. 53–4) in slightly different ways, is described below.

Let A, a matrix of order $n \times k$ and rank $r \leqslant \min(n, k)$, be partitioned as

$$A = \begin{bmatrix} A_{11} & A_{12} \\ A_{21} & A_{22} \end{bmatrix}, \qquad [4.10]$$

where A_{11} is $r \times r$ of rank r. Now, by Turing's theorem (e.g. see Fox, 1964, p. 33) A_{11} may be uniquely expressed as $A_{11} = LDU$ (provided that no leading minor is zero), where L and U are $r \times r$ unit lower and upper triangular matrices respectively, and D is a diagonal matrix.

Premultiplication of A by

$$P_1 = \begin{bmatrix} D^{-1}L^{-1} & 0 \\ -A_{21}A_{11}^{-1} & I_{n-r} \end{bmatrix} \qquad [4.11]$$

gives

$$P_1A = \begin{bmatrix} U & D^{-1}L^{-1}A_{12} \\ 0 & 0 \end{bmatrix}. \qquad [4.12]$$

Thus, by performing elementary row operations on A, it may be reduced to the form [4.12]. Now further row operations on [4.12] can be performed to reduce it to $\begin{bmatrix} I_r & R \\ 0 & 0 \end{bmatrix}$. Symbolically this is achieved by premultiplication of [4.12] by

$$P_2 = \begin{bmatrix} U^{-1} & 0 \\ 0 & I_{n-r} \end{bmatrix}. \qquad [4.13]$$

The whole operation may be portrayed as

$$\begin{bmatrix} A_{11}^{-1} & 0 \\ -A_{21}A_{11}^{-1} & I \end{bmatrix}\begin{bmatrix} A_{11} & A_{12} \\ A_{21} & A_{22} \end{bmatrix} = \begin{bmatrix} I & A_{11}^{-1}A_{12} \\ 0 & 0 \end{bmatrix},$$

or as $BA = H$, and can of course be carried out in a single stage. The division into two stages is to facilitate comparison of Rao's and Rohde's methods.

Rao (1965, p. 25) has shown that this operation yields a g_1-inverse of A. Consider, for example, the case $n \geqslant k$. Let A be augmented with $n - k$ columns of zeros, so that H is square, and denote the augmented matrix $[A \quad 0]$ by A^*. Clearly, $H^2 = H$ and $A^*BA^* = B^{-1}H^2 = B^{-1}H = A^*$. If B is partitioned as $B = \begin{bmatrix} B_1 \\ B_2 \end{bmatrix}$ to conform with the partitioning of A^*, then $A^*BA^* = A^*$ implies

$$[A \quad 0]\begin{bmatrix} B_1 \\ B_2 \end{bmatrix}[A \quad 0] = [AB_1A \quad 0],$$

i.e. $B_1 = A^{g_1}$. Hence, if elementary row-operations are performed on the augmented matrix $[A \quad I]$ until it is reduced to $[H \quad B]$, the first k rows of B will be a g_1-inverse of A, with full row-rank. In practice, of course, the operations do not start with the matrix in the form [4.10], but the process, by passing over zero pivots, equivalently produces the necessary changes of rows and columns.

A more direct verification of the result $B_1 = A^{g_1}$ is as follows. Let A be partitioned as in [4.10], but with $[A_{21} \quad A_{22}]$ further partitioned so that A is now of the form

$$A = \begin{bmatrix} A_{11} & A_{12} \\ A_{211} & A_{221} \\ A_{212} & A_{222} \end{bmatrix},$$

where A_{211} is $(k - r) \times r$. Now let B_1 be the $k \times n$ matrix

$$B_1 = \begin{bmatrix} A_{11}^{-1} & 0 & 0 \\ -A_{211}A_{11}^{-1} & I_{k-r} & 0 \end{bmatrix},$$

so that

$$B_1A = \begin{bmatrix} I & A_{11}^{-1}A_{12} \\ 0 & 0 \end{bmatrix}$$

since $\begin{bmatrix} A_{221} \\ A_{222} \end{bmatrix} - \begin{bmatrix} A_{211} \\ A_{212} \end{bmatrix} A_{11}^{-1}A_{12} = 0$. Clearly $AB_1A = A$. This implies that $n - k$ rows of A may be ignored in the sweep-out process, so that row operations are performed on the augmented matrix

$$\left[\begin{array}{cc|cc} A_{11} & A_{12} & I_r & 0 \\ A_{211} & A_{221} & 0 & I_{k-r} \end{array}\right]$$

to obtain

$$\left[\begin{array}{cc|c} I & R & \\ & & C \\ 0 & 0 & \end{array}\right],$$

in which case $[C \quad 0] = A^{g_1}$.

The situation where $n \leqslant k$ can be handled in a similar fashion, A being now augmented with $k - n$ rows of zeros.

Rohde, on the other hand, proposes to utilize both row and column operations in his method for finding A^{g_1}. The augmented matrix

$$\begin{bmatrix} A & I_n \\ I_k & 0 \end{bmatrix}$$

is reduced to

$$\begin{bmatrix} N & P_1 \\ P_2 & 0 \end{bmatrix},$$

where $N = P_1AP_2$ is of the form $\begin{bmatrix} I & K \\ 0 & 0 \end{bmatrix}$, and P_1 and P_2 are as in [4.11] and [4.13], except that the unit matrix in P_2 is of order $k - r$. The reduction may be represented in terms of partitioned matrices as follows:

$$\begin{bmatrix} P_1 & 0 \\ 0 & I_k \end{bmatrix} \begin{bmatrix} A & I_n \\ I_k & 0 \end{bmatrix} \begin{bmatrix} P_2 & 0 \\ 0 & I_n \end{bmatrix}$$

$$= \begin{bmatrix} P_1AP_2 & P_1 \\ P_2 & 0 \end{bmatrix}$$

$$= \left[\begin{array}{cc|c} I_r & D^{-1}L^{-1}A_{12} & \\ & & P_1 \\ 0 & 0 & \\ \hline \multicolumn{2}{c|}{P_2} & 0 \end{array}\right]$$

Now by Theorem 1.3, $A^{g_1} = P_2N^{g_1}P_1$, and convenient choices for N^{g_1} are provided by quasi-triangular matrices of the type $\begin{bmatrix} I & V \\ 0 & W \end{bmatrix}$, where V and W are arbitrary. The choice $N^{g_1} = \begin{bmatrix} I_r & 0 \\ 0 & 0 \end{bmatrix}'$ would be the simplest of course, but then $P_2 \begin{bmatrix} I_r & 0 \\ 0 & 0 \end{bmatrix}' P_1$ is actually a g_2-inverse of A. This matter will be further discussed in § 4.4.

It may be necessary in both Rao's and Rohde's methods to apply some permutation matrix in order to obtain a unit matrix in the leading position in the reduced matrix. However, this does not introduce any further problem since, if P is a permutation matrix, $(AP)^{g_1} \equiv P'A^{g_1}$. Furthermore, in Rohde's method it is unnecessary at each stage of the sweep-out to divide each pivotal row by the selected pivotal element. This would lead to a reduced matrix of the form

$$N = \begin{bmatrix} D_r & K \\ 0 & 0 \end{bmatrix} \qquad [4.14]$$

where D_r is an $r \times r$ nonsingular diagonal matrix. In this case the leading submatrix in N^{g_1} is D_r^{-1}.

Although Rao's method is more compact than Rohde's, the latter is nevertheless a more general procedure. It allows some freedom in the choice of N^{g_1} and, furthermore, it may be used to find symmetric g_1-inverses by means of a reduction of the type $PAP' = N$. If A is symmetric, then $A_{11} = LD^{\frac{1}{2}}D^{\frac{1}{2}}U = U'D^{\frac{1}{2}}D^{\frac{1}{2}}L'$, and therefore $L' = U$ since the decomposition is unique, i.e. $A_{11} = L_1L_1'$ where L_1 is positive lower triangular. The choice of the matrix P as

$$P = \begin{bmatrix} L_1^{-1} & 0 \\ -A_{21}A_{11}^{-1} & I \end{bmatrix}$$

gives $N = \begin{bmatrix} I & 0 \\ 0 & 0 \end{bmatrix}$, and by choosing a symmetric N^{g_1} a symmetric A^{g_1} is obtained.

While on the subject, it is perhaps worth noting that by applying Aitken's triple product method, it is possible to produce Rao's g_1-inverse, a g_1-inverse obtained by applying the equivalent column operations, and the inverse of A_{11} (and so a g_2-inverse of A) all in one process.

Expressions for g-inverses in terms of ordinary inverses are always useful. If S is a positive semidefinite matrix, then $(S + L'L)^{-1}$, where L is some matrix complementary to S, provides a compact method of finding a symmetric g_1-inverse of S (cf. Theorem 2.19). In statistical problems the composition of L is usually apparent from the nature of S.

The expressions for g_1-inverses of two-way partitioned matrices and partitioned positive semidefinite matrices, developed in § 3.2 and § 3.3 respectively, could also be used to calculate g_1-inverses, especially if a part of the matrix is in a form which readily admits a g_1-inverse.

4.4 g_2-inverses

As was mentioned in the previous section, the computation of A^{g_2} may be achieved by using the sweep-out process employed by Rohde for calculating A^{g_1}. Thus, for P_1 and P_2, as in his method,

$$A^{g_2} = P_2 N^{g_2} P_1 = P_2 \begin{bmatrix} I_r & 0 \\ 0 & 0 \end{bmatrix}' P_1 .$$

Actually, as is evident if $P_2 \begin{bmatrix} I_r & 0 \\ 0 & 0 \end{bmatrix}' P_1$ is multiplied out, this process

is merely a disguised way of obtaining

$$A^{g_2} = \begin{bmatrix} A_{11}^{-1} & 0 \\ 0 & 0 \end{bmatrix}, \qquad [4.15]$$

the g_2-inverse mentioned in § 1.5. The expression [4.15] would appear to be the simplest and most compact formula for calculating A^{g_2}, or for that matter, by virtue of the inclusiveness of the definition of A^{g_1}, a g_1-inverse of A. In many statistical problems it is often possible to select A_{11} so that it is diagonal or very nearly diagonal. This method would then definitely lead to the least arithmetic for the data at hand. If A is symmetric then [4.15] will, of course, give a symmetric A^{g_2}.

It is interesting at this stage to return to the two general forms for a solution to equations $A\mathbf{x} = \mathbf{h}$ presented in [1.17] and [1.19]. Searle (1966) expounds on the computational advantages of the form [1.17] over that presented in [1.19]. However, it is evident from [4.15] that, although [1.17] is simpler both theoretically and notationally, it can possess no computational advantages whatsoever. In fact, the form [1.19] is also one of notational convenience, and the calculation would not necessarily involve a re-ordering of the rows and columns.

The computations involved in the sweep-out method can be further reduced by not dividing each pivotal row by the selected pivotal element. The reduced matrix will be of the form [4.14], in which case

$$A^{g_2} = P_2 \begin{bmatrix} D_r^{-1} & 0 \\ 0 & 0 \end{bmatrix}' P_1$$

as in [1.22].

4.5 g_3-inverses

Since any matrix of the form $G = (A'A)^{g_1}A'$ is a g_3-inverse of A, the computation of a g_3-inverse can be reduced to that of finding a g_1-inverse. However, in view of the remarks made in the preceding section, it is simpler to compute a g_2-inverse of $A'A$. This does not lead to any further particularization of G (cf. Theorem 2.13, Corollary 1).

If A is partitioned as $A = [U \quad V]$, where the columns of U form a basis for the column-space of A, then, by [4.15],

$$A^{g_3} = (A'A)^{g_2}A' = \begin{bmatrix} (U'U)^{-1} & 0 \\ 0 & 0 \end{bmatrix} \begin{bmatrix} U' \\ V' \end{bmatrix} = \begin{bmatrix} U^g \\ 0 \end{bmatrix}.$$

This result was obtained through a different approach in Theorem 3.3, Corollary.

4.6 The Cholesky technique

It was mentioned in § 4.2 that the Doolittle method, well known as a computational method of regular matrix inversion and of solving linear equations of full rank, may be adapted to handle g-inverse computations.

The Cholesky (or square-root) method, although less well known in the U.S.A. perhaps, is also widely known among statisticians as a computationally accurate method for inverting matrices and solving linear equations of full rank. It will now be shown that it can be used for the same purposes in conjunction with g-inverses as the Doolittle.

In the case of a symmetric matrix A of full rank the Cholesky method is based on the resolution $A = LL'$, where L is a positive lower triangular matrix, which is real only if A is positive definite. The matrix L is obtained in a systematic way from A, and L^{-1} is computed recursively from L. Finally $A^{-1} = (L')^{-1}L^{-1}$.

If A is singular, but still symmetric, the same process will yield a g_2-inverse of A. Let A, of order k and rank $r \leqslant k$, be partitioned as

$$A = \begin{bmatrix} A_{11} & A_{12} \\ A_{21} & A_{22} \end{bmatrix},$$

where A_{11} is $r \times r$ nonsingular. The Cholesky technique may be used to calculate A_{11}^{-1}, and then, by [4.15], A^{g_2}. Rao (1955), in considering the solution of singular least-squares equations, used the Cholesky technique to calculate a "pseudo-inverse". His methods led to the

g_2-inverse obtained above.

Rao also mentioned that in practice it is not necessary to recognize the dependent rows of A in advance. They will be discovered during the process of computation, since, if the jth row of A is a linear combination of the first $j - 1$ rows, then the jth diagonal element of $L(A = LL')$ will be zero, in which case the entire row of L and the corresponding column are omitted. The reduced triangular matrix will be the one required for further inversion. A g_2-inverse of A is therefore readily obtained by means of the Cholesky method. Healy (1968b, c) has presented computer programs for the calculation of L and A_{11}^{-1} for the case when A is positive semidefinite and partitioned as above.

An adaptation of the Cholesky technique which permits the calculation of a g_1-inverse of a positive semidefinite matrix has been mentioned by Rao (1965, p. 59). We consider here, however, the case A symmetric. Let A be partitioned as above and let A_{11} be factored as L_1L_1', as in the Cholesky method. Furthermore, A_{12} and A_{22} are expressible as $A_{12} = A_{11}R$ and $A_{22} = R'A_{11}R$, for some $r \times (k - r)$ matrix R. Now let T denote the positive lower triangular matrix

$$T = \begin{bmatrix} L_1^{-1} & 0 \\ -R' & I \end{bmatrix}.$$

Clearly $TA = L'$, where

$$L' = \begin{bmatrix} L_1' & L_1'R \\ 0 & 0 \end{bmatrix}.$$

Also, $LL' = A$, whence $A = LL' = AT'TA$, i.e. $T'T = A^{g_1}$. If the square-root technique is applied to A, with the convention that, if any diagonal element in the factorization is zero (in which case the entire row is also zero) it is replaced by unity, it follows that a matrix of the form

$$(L^*)' = \begin{bmatrix} 0 & 0 \\ 0 & I_{k-r} \end{bmatrix} + L'$$

will be obtained. It can easily be shown that $(L^*)^{-1} = T$, so that the problem of determining a g_1-inverse of A reduces to that of determining L'. Rao, in fact, suggests that T be computed directly by applying the square-root technique to the augmented matrix $[A \quad I]$, so that the resultant augmented matrix will be of the form $[(L^*)' \quad T]$. The same convention is, of course, adopted when a zero diagonal element is encountered. However, it would appear that in the case of large matrices, the computation of T by the direct inversion of L^* has an advantage of compactness over Rao's augmented scheme.

Chapter 5

SINGULAR NORMAL VARIATES

5.1 Conditional means and variances of the normal multivariate distribution

Let $\mathbf{x}$ be a k-variate normal vector with mean $\boldsymbol{\mu}$ and variance matrix V, i.e. $N(\boldsymbol{\mu}, V)$. Let $\mathbf{x}$ be partitioned as $\begin{bmatrix} \mathbf{x}_1 \\ \mathbf{x}_2 \end{bmatrix}$, where $\mathbf{x}_2$ corresponds to the last s elements of $\mathbf{x}$, and let $\boldsymbol{\mu}$ and V be conformably partitioned as $\begin{bmatrix} \boldsymbol{\mu}_1 \\ \boldsymbol{\mu}_2 \end{bmatrix}$

and

$$\begin{bmatrix} V_{11} & V_{12} \\ V_{21} & V_{22} \end{bmatrix}. \qquad [5.1]$$

We consider the case where V_{22} is singular with rank $p < s$. Marsaglia (1964), Rao (1965, pp. 441–2), and Harris & Helvig (1966) have discussed the derivation of expressions for the conditional mean ($\boldsymbol{\mu}_c$) and variance matrix (V_c) of $\mathbf{x}_1$ for fixed $\mathbf{x}_2$ in terms of g-inverses of V_{22}. However, as Harris & Helvig point out, Marsaglia did not establish the invariance of his expressions for $\boldsymbol{\mu}_c$ and V_c, and the same applies to Rao's derivation. The following derivation is in many respects similar to that of Harris & Helvig, except that a g_1-inverse of V_{22} will be used instead of V_{22}^g.

Since V is positive semidefinite it is expressible as $V = K'K$, where K may be conformably partitioned so that

$$K'K = \begin{bmatrix} K_1'K_1 & K_1'K_2 \\ K_2'K_1 & K_2'K_2 \end{bmatrix}.$$

Now, since $r(K_2'K_2) = p < s$, there exists a matrix H, of full row-rank $s - p$, such that

$$HK_2' = 0. \qquad [5.2]$$

It therefore follows, from [5.2], that

$$HV_{22}H' = \mathrm{E}[H(\mathbf{x}_2 - \boldsymbol{\mu}_2)(\mathbf{x}_2 - \boldsymbol{\mu}_2)'H'] = 0$$

and thus $s - p$ linearly independent linear relations $H(\mathbf{x}_2 - \boldsymbol{\mu}_2) = 0$ hold with probability one. This implies, from [5.2], that $\mathbf{x}_2 - \boldsymbol{\mu}_2$ is contained in the column-space of K_2', i.e. with probability one,

$$\mathbf{x}_2 - \boldsymbol{\mu}_2 = K_2'\mathbf{z} \qquad [5.3]$$

for some vector $\mathbf{z}$. The qualifying phrase "with probability one" will henceforth be implied, and will not be included at each appropriate stage.

Following Harris & Helvig, the conditional distribution is obtained by seeking a transformation of the variables

$$\mathbf{y} = \begin{bmatrix} \mathbf{y}_1 \\ \mathbf{y}_2 \end{bmatrix} = P\mathbf{x} = \begin{bmatrix} I & -C \\ 0 & I \end{bmatrix}\begin{bmatrix} \mathbf{x}_1 \\ \mathbf{x}_2 \end{bmatrix}$$

such that $\operatorname{cov}(\mathbf{y}_1, \mathbf{y}_2) = 0$. Then, since $\mathbf{y}_1$ and $\mathbf{y}_2$ are independent, the conditional distribution of $\mathbf{y}_1$ given $\mathbf{y}_2$ is the same as the marginal distribution of $\mathbf{y}_1$.

Now $\operatorname{cov}(\mathbf{y}_1, \mathbf{y}_2) = [I \;\; -C]\, V \begin{bmatrix} 0 \\ I \end{bmatrix} = V_{12} - CV_{22}$. The matrix C is therefore required to satisfy

$$CV_{22} = V_{12}. \qquad [5.4]$$

Clearly, since $V_{22} = K_2'K_2$ and $V_{12} = K_1'K_2$, V_{12} is contained in the row-space of V_{22}, i.e. equations [5.4] are consistent (cf. Theorem 1.6, Corollary 2) and C exists.

It now follows that the expectation of $\mathbf{x}_1$ for fixed $\mathbf{x}_2$ is

$$\boldsymbol{\mu}_c = \boldsymbol{\mu}_1 + C(\mathbf{x}_2 - \boldsymbol{\mu}_2), \qquad [5.5]$$

and by [5.4] the variance matrix of the conditional distribution is

$$V_c = V_{11} - CV_{21}. \qquad [5.6]$$

Harris & Helvig point out that Marsaglia chose the particular solution $C = V_{12}V_{22}^{g}$. Similarly, Rao used the particular set of solutions $C = V_{12}G$, where $G = V_{22}^{g_1}$. However, before the choice of a particular solution is justified, it is necessary to establish the invariance of $\boldsymbol{\mu}_c$ and V_c under the many choices of solutions to [5.4]. The invariance is shown as follows:

By Theorem 1.6, Corollary 3, the general solution to [5.4] is

$$C = V_{12}V_{22}^{g_1} + Y(I - V_{22}V_{22}^{g_1}), \qquad [5.7]$$

where Y is arbitrary. Postmultiplication of [5.7] by V_{21} gives, by [2.6], $CV_{21} = V_{12}V_{22}^{g_1}V_{21}$, and it follows from Theorem 2.12 that this is unique. Thus [5.6] becomes

$$V_c = V_{11} - V_{12}V_{22}^{g_1}V_{21}. \qquad [5.8]$$

Likewise, the substitution of [5.3] in [5.5] gives

$$\boldsymbol{\mu}_c = \boldsymbol{\mu}_1 + CK_2'\mathbf{z},$$

and the uniqueness of CK_2' follows immediately from [5.7] and [2.6]. Therefore

$$\boldsymbol{\mu}_c = \boldsymbol{\mu}_1 + V_{12}V_{22}^{g_1}(\mathbf{x}_2 - \boldsymbol{\mu}_2). \qquad [5.9]$$

It can now be seen that the expressions for the conditional expectation and variance obtained by Marsaglia and Rao are actually correct. The invariance of $\boldsymbol{\mu}_c$ and V_c implies, of course, that $V_{22}^{g_1}$ may be replaced by V_{22}^{g}. However, we retain $V_{22}^{g_1}$ to emphasize the fact that only the single condition is required.

In the following theorem, which generalizes a familiar result in the theory of the nonsingular normal multivariate distribution, $\mathbf{x}_1$ becomes the single variate x_1 and $\mathbf{x}_2$ the remaining $k - 1$ variates.

THEOREM 5.1 If v_{11}^* denotes the variance of x_1 in the conditional distribution of x_1 for fixed $x_2, x_3, \ldots, x_k$, then $v_{11}^* \leqslant v_{11}$.

Proof: By definition,

$$v_{11}^* = v_{11} - V_{12}V_{22}^{g_1}V_{21}.$$

Now, since V is expressible as $K'K$, it follows from Theorem 2.12 that

$$V_{12}V_{22}^{g_1}V_{21} = V_{12}V_{22}^{g}V_{21} = (K_1'K_2K_2^{g})(K_1'K_2K_2^{g})',$$

i.e. $V_{12}V_{22}^{g_1}V_{21}$ is positive semidefinite and the result follows.

5.2 Regression properties of the multivariate normal distribution

One of the first methods of multivariate analysis with which the statistician becomes acquainted is that of partial correlation analysis. Partial correlation coefficients, multiple correlation coefficients and partial regression coefficients are readily described in the case of a full-rank normal multivariate distribution; see, for example, Anderson

(1958, pp. 27–34).

In this section we tentatively investigate the regression properties of a singular multivariate normal distribution. In practice, of course, one does not usually worry about the case V singular, since the problem may be reduced to the nonsingular case by simply dropping the appropriate number of redundant variates. In certain cases, however, it may be convenient to have general methods of calculating partial regression and correlation coefficients whether V is singular or not, especially, for example, if the redundancies are not immediately obvious.

The multiple correlation coefficient

The multiple correlation coefficient $R_{1,2\ldots k}$ may be defined as the correlation coefficient between the two variates (scalar) x_1 and $\hat{x}_1$, where

$$\hat{x}_1 = \mu_1 + V_{12}V_{22}^{g_1}(\mathbf{x}_2 - \boldsymbol{\mu}_2).$$

Thus

$$R_{1,2\ldots k} = \frac{\mathrm{E}\{(x_1 - \mu_1)[\hat{x}_1 - \mathrm{E}(\hat{x}_1)]\}}{\{\mathrm{E}(x_1 - \mu_1)^2\,\mathrm{E}[\hat{x}_1 - \mathrm{E}(\hat{x}_1)]^2\}^{\frac{1}{2}}}. \qquad [5.10]$$

Now the numerator in [5.10] may be written as

$$\begin{aligned}\mathrm{E}[(x_1 - \mu_1)V_{12}V_{22}^{g_1}(\mathbf{x}_2 - \boldsymbol{\mu}_2)] &= \mathrm{E}[(x_1 - \mu_1)(\mathbf{x}_2 - \boldsymbol{\mu}_2)'(V_{22}^{g_1})'V_{21}] \\ &= V_{12}(V_{22}^{g_1})'V_{21} \\ &= V_{12}V_{22}^{g_1}V_{21} \quad \text{by Theorem 2.12.}\end{aligned}$$

Furthermore,

$$\begin{aligned}\mathrm{E}[\hat{x}_1 - \mathrm{E}(\hat{x}_1)]^2 &= \mathrm{E}[V_{12}V_{22}^{g_1}(\mathbf{x}_2 - \boldsymbol{\mu}_2)(\mathbf{x}_2 - \boldsymbol{\mu}_2)'(V_{22}^{g_1})'V_{21}] \\ &= V_{12}V_{22}^{g_1}V_{22}(V_{22}^{g_1})'V_{21} \\ &= V_{12}V_{22}^{g_1}V_{21} \quad \text{by [2.6].}\end{aligned}$$

Thus,

$$R_{1,2\ldots k} = \left(\frac{V_{12}V_{22}^{g_1}V_{21}}{v_{11}}\right)^{\frac{1}{2}}. \qquad [5.11]$$

In view of the discussion in § 5.2, it is evident that the multiple correlation coefficient is uniquely determined.

It is interesting to note that the "generalized correlation matrix" defined, for example by Khatri (1964) may be generalized to include the case V singular. If V is partitioned as in [5.1], then, for V

positive definite, the generalized correlation matrix R^* is defined as

$$R^* = V_{11}^{-\frac{1}{2}} V_{12} V_{22}^{-1} V_{21} V_{11}^{-\frac{1}{2}}.$$

When V is positive semidefinite, R^* can be defined as

$$R^* = (V_{11}^{\frac{1}{2}})^g V_{12} V_{22}^{g_1} V_{21} (V_{11}^{\frac{1}{2}})^g. \qquad [5.12]$$

The expression [5.12] is unique since the square root of a positive semidefinite matrix is unique (Rao, 1965, p. 55) and $V_{12} V_{22}^{g_1} V_{21}$ is invariant under the choice of g-inverse of V_{22}. For the particular case where $\mathbf{x}_1$ consists of a single element only, R^* reduces to $R_{1,2\ldots k}$, as mentioned by Khatri.

Regression equations

Assume once again that $\mathbf{x}$, $\boldsymbol{\mu}$ and V are partitioned as in §5.1. Then the loci of the means of the $\mathbf{x}_1$ arrays are given by

$$\boldsymbol{\mu}_{\mathbf{x}_1, \mathbf{x}_2} = \boldsymbol{\mu}_1 + V_{12} V_{22}^{g_1} (\mathbf{x}_2 - \boldsymbol{\mu}_2),$$

which are the regression equations of $\mathbf{x}_1$ on $\mathbf{x}_2$. When V is positive definite, the matrix $V_{12} V_{22}^{-1}$ is defined as the matrix of partial regression coefficients. However, when V is positive semidefinite it would appear that there is little point in defining an algebraically unique matrix of partial regression coefficients, $V_{12} V_{22}^{g}$, since this would still be arbitrary in the statistical sense. For, if H is some matrix such that $H(\mathbf{x}_2 - \boldsymbol{\mu}_2) = \mathbf{0}$, as in §5.1, then

$$\begin{aligned} \boldsymbol{\mu}_{\mathbf{x}_1, \mathbf{x}_2} &= \boldsymbol{\mu}_1 + (V_{12} V_{22}^{g} + \alpha H)(\mathbf{x}_2 - \boldsymbol{\mu}_2) \\ &= \boldsymbol{\mu}_1 - (V_{12} V_{22}^{g} + \alpha H)\boldsymbol{\mu}_2 + (V_{12} V_{22}^{g} + \alpha H)\mathbf{x}_2, \end{aligned}$$

where α is any scalar, i.e. αH may be added to the matrix of "regression coefficients" by subtracting $\alpha H \boldsymbol{\mu}_2$ from the constant term. Thus, if $V_{12} V_{22}^{g} = C$, the interpretation of any c_{ij} in the usual manner will not be possible.

As regards partial correlation coefficients in the singular case, there is also some difficulty due to the possible occurrence of zero conditional variances. It would be perhaps feasible to define a partial correlation coefficient as in the full-rank case, with the convention that the parameter be undefined for any pair of variates whenever either of the appropriate diagonal elements of V_c is zero.

5.3 Reversible transformations of singular to nonsingular variates

Let $\mathbf{x}$ be a k-variate vector with mean $\boldsymbol{\mu}$ and variance matrix V, of rank r. If $r = k$, it is possible to express $\mathbf{x}$ in terms of k variates $\mathbf{y}$

with zero mean and unit variance by using the transformation $\mathbf{x} = \boldsymbol{\mu} + V^{\frac{1}{2}}\mathbf{y}$. Since $V^{\frac{1}{2}}$ is positive definite, the transformation is reversible. It will be demonstrated in this section that g-inverses may be used to derive similar reversible transformations for the case $r < k$. Such a transformation has been derived by Rayner & Livingstone (1965), who based their result on the reduction of V to leading diagonal canonical form.

The transformation used by Rayner & Livingstone is, however, a special case of that put forward by Anderson (1958, pp. 25–26): If $\mathbf{x}$ is arranged so that $\mathbf{x}_1$ in the partitioning $\begin{bmatrix} \mathbf{x}_1 \\ \mathbf{x}_2 \end{bmatrix}$ has full rank r, then there exists a nonsingular matrix P such that $PVP' = \begin{bmatrix} I_r & 0 \\ 0 & 0 \end{bmatrix}$. Hence variates $\mathbf{y} = P\mathbf{x}$ have variance matrix $\begin{bmatrix} I_r & 0 \\ 0 & 0 \end{bmatrix}$, and since $\mathbf{x} = P^{-1}\mathbf{y}$ and $\text{var}(\mathbf{x}) = P^{-1}\begin{bmatrix} I_r & 0 \\ 0 & 0 \end{bmatrix}(P^{-1})' = V$, the transformation is reversible. The following may also be noted:

(a) Since $\text{var}(\mathbf{y}_2) = 0$, where $\mathbf{y}_2$ consists of the last $k - r$ elements of $\mathbf{y}$, $\mathbf{y}_2 = \text{E}(\mathbf{y}_2)$ with probability one.

(b) $V = P^{-1}\begin{bmatrix} I_r & 0 \\ 0 & 0 \end{bmatrix}(P^{-1})'P'PP^{-1}\begin{bmatrix} I_r & 0 \\ 0 & 0 \end{bmatrix}(P^{-1})' = VP'PV$, i.e. $P'P$ is a full-rank g_1-inverse of V.

(c) Let P^{-1} be partitioned as $P^{-1} = [K \quad B]$; then $\mathbf{x} = P^{-1}\mathbf{y}$ may be written in partitioned form as

$$\mathbf{x} = [K \quad B]\begin{bmatrix} \mathbf{y}_1 \\ \mathbf{y}_2 \end{bmatrix} = K\mathbf{y}_1 + B\mathbf{y}_2,$$

i.e. the transformation is of the type $\mathbf{x} = K\mathbf{y}_1 + \textit{constant}$, with

$$[K \quad B]\begin{bmatrix} I_r & 0 \\ 0 & 0 \end{bmatrix}\begin{bmatrix} K' \\ B' \end{bmatrix} = KK' = V. \qquad [5.13]$$

Rayner & Livingstone, in effect, took $P = \begin{bmatrix} \Lambda^{-\frac{1}{2}} & 0 \\ 0 & A \end{bmatrix} H = \begin{bmatrix} \Lambda^{-\frac{1}{2}} H_1 \\ A H_2 \end{bmatrix}$, where

H is an orthogonal matrix such that $HVH' = \begin{bmatrix} \Lambda & 0 \\ 0 & 0 \end{bmatrix}$, Λ is a diagonal

matrix of nonzero latent roots of V, and A is an arbitrary nonsingular

matrix. Thus $P^{-1} = H' \begin{bmatrix} \Lambda^{\frac{1}{2}} & 0 \\ 0 & A^{-1} \end{bmatrix}$, and $K = H_1' \Lambda^{\frac{1}{2}}$. Actually $\begin{bmatrix} \Lambda^{-\frac{1}{2}} & 0 \\ 0 & A \end{bmatrix}$

is a full-rank g_1-inverse of $HV^{\frac{1}{2}}H'$ which also satisfies conditions (3)

and (4), and in the same way as in Theorem 2.16, Corollary 4, $H' \begin{bmatrix} \Lambda^{-\frac{1}{2}} & 0 \\ 0 & A \end{bmatrix} H$

is a full-rank g_1-inverse of $V^{\frac{1}{2}}$ which also satisfies conditions (3) and (4), i.e. $P = H(V^{\frac{1}{2}})^{g_1}$, a form which symbolizes the approach of these authors.

As a second example, P may be taken as

$$\begin{bmatrix} V_{11}^{-\frac{1}{2}} & 0 \\ -V_{21} V_{11}^{-1} & I \end{bmatrix}, \qquad [5.14]$$

where the partitioning of V conforms with that of $\mathbf{x}$. This may be expressed as

$$\begin{bmatrix} V_{11}^{-\frac{1}{2}} & 0 \\ 0 & A \end{bmatrix} \begin{bmatrix} I & 0 \\ -V_{21} V_{11}^{-1} & I \end{bmatrix},$$

which makes it clear that this transformation is similar in all respects to the foregoing.

Rayner & Livingstone also showed that, if P were taken as $H(V^{\frac{1}{2}})^{g} = \Lambda^{-\frac{1}{2}} H_1$ of order $r \times k$ (i.e. $A = 0$), reversibility is retained, provided that $H_2 \boldsymbol{\mu} = \mathbf{0}$ and in particular if $\boldsymbol{\mu} = \mathbf{0}$. Hence for the case $\boldsymbol{\mu} = \mathbf{0}$ they proposed the transformation $\mathbf{x} = H_1' \Lambda^{\frac{1}{2}} \mathbf{y} = K\mathbf{y}$ and $\mathbf{y} = \Lambda^{-\frac{1}{2}} H_1 \mathbf{x}$, where $\mathbf{y}$ now has r elements. This may be generalized and used to provide an alternative method for the case $\boldsymbol{\mu} \neq \mathbf{0}$ as follows.

Since V is positive semidefinite, of rank r, it may be factored as $V = KK'$, where K is $k \times r$ of rank r. We note also, by arguments similar to those leading up to [5.3], that $\mathbf{x} - \boldsymbol{\mu}$ is contained in the column-space of V (with probability one), i.e.

$$\mathbf{x} - \boldsymbol{\mu} = V\mathbf{z} \qquad [5.15]$$

for some vector $\mathbf{z}$. Consider now variates

$$\mathbf{y} = L(\mathbf{x} - \boldsymbol{\mu}), \qquad [5.16]$$

where L, of order $r \times k$, is any left inverse of K. It follows from [5.16] that $\mathbf{y}$, an r-dimensional vector, has zero expectation and that $\mathrm{var}(\mathbf{y}) = LVL' = LKK'L' = I_r$. The reversibility of the transformation is shown by premultiplying [5.16] by K. Then $K\mathbf{y} = KL(\mathbf{x} - \boldsymbol{\mu}) = KLV\mathbf{z}$ (by [5.15]) $= KLKK'\mathbf{z} = KK'\mathbf{z} = V\mathbf{z} = \mathbf{x} - \boldsymbol{\mu}$,

i.e.

$$\mathbf{x} = \boldsymbol{\mu} + K\mathbf{y}. \qquad [5.17]$$

Clearly, $\mathrm{E}(\mathbf{x}) = \boldsymbol{\mu}$ and $\mathrm{var}(\mathbf{x}) = KK' = V$. The matrix K must be as given in [5.13], while L may be taken as $K^{g_3^*} = K'(KK')^{g_1} = K'V^{g_1}$ or $K^{g} = (K'K)^{-1}K'$.

5.4 Conditions for a second-degree polynomial in normal variates to have noncentral χ^2 distribution

Conditions under which a quadratic polynomial $\mathbf{x}'Q\mathbf{x} + \mathbf{m}'\mathbf{x} + d$, where $\mathbf{x}$ is normal multivariate, has noncentral χ^2 distribution are given in the following theorems:

THEOREM 5.2 (Khatri, 1962) If $\mathbf{x}$ is $\mathrm{N}(\boldsymbol{\mu}, I)$, a set of necessary and sufficient conditions for $\mathbf{x}'Q\mathbf{x} + \mathbf{m}'\mathbf{x} + d$ to have noncentral χ^2 distribution is $Q^2 = Q$, $\mathbf{m}' = \mathbf{m}'Q$, and $d = \frac{1}{4}\mathbf{m}'\mathbf{m}$, the degrees of freedom and noncentrality parameter being given by $f = \mathrm{tr}(Q)$ and $\lambda = d$, respectively.

Notice that these conditions do not involve $\boldsymbol{\mu}$.

THEOREM 5.3 (Khatri, 1963) If $\mathbf{x}$ is $\mathrm{N}(\boldsymbol{\mu}, V)$, a set of necessary and sufficient conditions for $\mathbf{x}'Q\mathbf{x} + \mathbf{m}'\mathbf{x} + d$ to have noncentral χ^2 distribution is:

(a) $VQVQV = VQV$,

(b) $(Q\boldsymbol{\mu} + \frac{1}{2}\mathbf{m})'V = (Q\boldsymbol{\mu} + \frac{1}{2}\mathbf{m})'VQV$,

(c) $\boldsymbol{\mu}'Q\boldsymbol{\mu} + \mathbf{m}'\boldsymbol{\mu} + d = (Q\boldsymbol{\mu} + \frac{1}{2}\mathbf{m})'V(Q\boldsymbol{\mu} + \frac{1}{2}\mathbf{m})$.

Rayner & Livingstone (1965) used a transformation similar to the first of the two transformations of § 5.3 to prove Theorem 5.3 as an extension of Theorem 5.2. However, their proof may be considerably simplified by using transformation [5.17].

Under this transformation $\mathbf{x}'Q\mathbf{x} + \mathbf{m}'\mathbf{x} + d$ becomes

$$\mathbf{y}'K'QK\mathbf{y} + (2\boldsymbol{\mu}'Q + \mathbf{m}')K\mathbf{y} + (\boldsymbol{\mu}'Q\boldsymbol{\mu} + \mathbf{m}'\boldsymbol{\mu} + d),$$

a quadratic polynomial in variates $\mathbf{y}$, $N(\mathbf{0}, I)$. Application of Theorem 5.2 yields the three conditions:

(a) $K'QVQK = K'QK$

(b) $(Q\mu + \frac{1}{2}\mathbf{m})'K = (Q\mu + \frac{1}{2}\mathbf{m})'VQK$

(c) $\mu'Q\mu + \mathbf{m}'\mu + d = (Q\mu + \frac{1}{2}\mathbf{m})'V(Q\mu + \frac{1}{2}\mathbf{m})$.

By the lemma mentioned in Theorem 2.12, condition (a) is equivalent to $VQVQV = VQV$. Similarly (b) is equivalent to $(Q\mu + \frac{1}{2}\mathbf{m})'V = (Q\mu + \frac{1}{2}\mathbf{m})'VQV$, and the result has been proved.

The degrees of freedom are

$$f = \operatorname{tr}(K'QK) = \operatorname{tr}(QKK') \quad \text{or} \quad \operatorname{tr}(KK'Q)$$
$$= \operatorname{tr}(QV) \quad \text{or} \quad \operatorname{tr}(VQ),$$

and the noncentrality parameter is

$$\lambda = \mu'Q\mu + \mathbf{m}'\mu + d = (Q\mu + \tfrac{1}{2}\mathbf{m})'V(Q\mu + \tfrac{1}{2}\mathbf{m}).$$

Mitra (1968a) has studied the representation of the condition $VQVQV = VQV$ in terms of g_1-inverses.

Corollary 1 It is evident from the form of the noncentrality parameter that the necessary and sufficient conditions for a central χ^2 distribution are $VQVQV = VQV$, $(Q\mu + \frac{1}{2}\mathbf{m})'V = \mathbf{0}$, and $\mu'Q\mu + \mathbf{m}'\mu + d = 0$.

The second of these three conditions was incorrectly given by Rao (1966) as $\mu'QV = \mathbf{0}$.

Corollary 2 If $\mathbf{x}$ is $N(\mathbf{0}, V)$, a sufficient condition for $\mathbf{x}'Q\mathbf{x}$ to have χ^2 distribution is $V = Q^{g_1}$ or $Q = V^{g_1}$. This result is evident from the condition $VQVQV = VQV$.

If V is positive definite, it follows that the condition $V = Q^{g_1}$ is both necessary and sufficient for χ^2 distribution, whereas $Q = V^{g_1}$ is necessary and sufficient if Q is positive definite (Shanbhag, 1968).

If the degrees of freedom of the distribution are specified in advance to be $r = \operatorname{r}(V)$, then the following result due to Khatri (1968) may be stated. The proof is as given by Khatri in a personal communication some years earlier.

Corollary 3 If $\mathbf{x}$ is $N(\mathbf{0}, V)$, a necessary and sufficient condition for $\mathbf{x}'Q\mathbf{x}$ to follow a χ^2 distribution with $r = \operatorname{r}(V)$ degrees of freedom is $Q = V^{g_1}$.

Proof: By means of the transformation [5.17] with $\boldsymbol{\mu} = \mathbf{0}$, $\mathbf{x}$ may be expressed as $\mathbf{x} = K\mathbf{y}$, where K, a $k \times r$ matrix of rank r, is such that $KK' = V$, and $\mathbf{y}$ is $\mathrm{N}(\mathbf{0}, I_r)$. Thus $\mathbf{x}'Q\mathbf{x} = \mathbf{y}'K'QK\mathbf{y}$. Now Graybill (1961, p. 82), for example, has shown that a necessary and sufficient condition for a quadratic form in variates $\mathrm{N}(\mathbf{0}, I)$ to have χ^2 distribution with f degrees of freedom is that the matrix of the quadratic form should be idempotent of rank f. Application of this condition to the quadratic form $\mathbf{y}'K'QK\mathbf{y}$ gives $K'QK = I_r$, which is equivalent to $KK'QKK' = KK'$ or $VQV = V$.

It therefore follows that the sufficiency condition $Q = V^{g_2}$ for $\mathbf{x}'Q\mathbf{x}$ to follow a χ^2 distribution with r degrees of freedom, given by Zelen & Federer (1965), is unnecessarily restrictive, as is also mentioned by Khatri (1968).

Chapter 6

THE LINEAR MODEL OF LESS THAN FULL RANK

6.1 Introduction

The linear model will be considered in the form

$$\mathbf{y} = X\boldsymbol{\beta} + \boldsymbol{\varepsilon}, \tag{6.1}$$

where $\mathbf{y}$ is an $n \times 1$ vector of observations, X is an $n \times k$ matrix, of rank $p < k$, of known constants, $\boldsymbol{\beta}$ is a $k \times 1$ vector of parameters, and $\boldsymbol{\varepsilon}$ is an $n \times 1$ vector of errors with zero expectation and variance matrix $\mathrm{E}(\boldsymbol{\varepsilon\varepsilon}') = \mathrm{var}(\mathbf{y}) = \sigma^2 I$. The case in which the observations have variance matrix V, possibly singular, is considered in the next chapter.

It is well known that under model [6.1] there are no unbiased estimates of $\boldsymbol{\beta}$, and with experimental design models two ways of handling this situation have been developed.

First, one may characterize the class of all linear functions $\boldsymbol{\theta}'\boldsymbol{\beta}$ for which unbiased linear estimates $\mathbf{a}'\mathbf{y}$ exist. These are the so-called "estimable functions" (Bose, 1944), which are characterized by the property that a solution to the equations

$$\mathbf{a}'X = \boldsymbol{\theta}' \tag{6.2}$$

exists, i.e. $\boldsymbol{\theta}'$ belongs to the p-dimensional row-space of X. By Theorem 1.5, Corollary, [6.2] is equivalent to

$$\boldsymbol{\theta}' = \boldsymbol{\theta}' X^{g_1} X, \tag{6.3}$$

and, since the row-spaces of X and $S = X'X$ are the same, to

$$\boldsymbol{\theta}' S^{g_1} S = \boldsymbol{\theta}'. \tag{6.4}$$

The second approach consists of specifying a set of q linear restrictions $L\boldsymbol{\beta} = \mathbf{c}$, where L is complementary to X. This leads to an augmented full-rank system of equations which admit a unique solution, a conditionally unbiased estimate subject to $L\boldsymbol{\beta} = \mathbf{c}$. This latter approach derives from arguments put forward by Yates & Hale (1939), Rao (1946), and Plackett (1950). In terms of the first approach the linear restrictions comprise a set of non-estimable functions of $\boldsymbol{\beta}$.

The concept of a generalized matrix inverse and its properties in the solution of linear equations are particularly applicable to the theory of the linear model of less than full rank. In recent years g-inverses have been used to restate existing results on estimability

(Bose, 1959; Rao, 1962, 1965; Goldman & Zelen, 1964; Searle, 1966) and on constrained linear models (Chipman, 1964; Goldman & Zelen; John, 1964; Rayner & Pringle, 1967).

The use of g-inverses in formulating estimation procedures for the model [6.1], together with certain related topics such as the fitting of augmented models and the analysis of multiple covariance, will be discussed in the ensuing sections.

6.2 Estimable linear functions

The problem of obtaining a *best* (*minimum variance*) *linear unbiased estimate* ("BLU estimate") of a linear function $\boldsymbol{\theta}'\boldsymbol{\beta}$ may be formulated as follows. A linear estimator $\hat{\boldsymbol{\beta}} = G\mathbf{y} + \mathbf{d}$ of $\boldsymbol{\beta}$ is sought such that $\boldsymbol{\theta}'\hat{\boldsymbol{\beta}}$ is a BLU estimate of $\boldsymbol{\theta}'\boldsymbol{\beta}$. The following theorem expresses the criteria for unbiasedness:

THEOREM 6.1 If $\hat{\boldsymbol{\beta}} = G\mathbf{y} + \mathbf{d}$, then $\boldsymbol{\theta}'\hat{\boldsymbol{\beta}}$ is an unbiased estimate of $\boldsymbol{\theta}'\boldsymbol{\beta}$ if and only if $G = X^{g_1}$ and $X\mathbf{d} = \mathbf{0}$.

Proof: For $\boldsymbol{\theta}'\hat{\boldsymbol{\beta}}$ to be an unbiased estimate of $\boldsymbol{\theta}'\boldsymbol{\beta}$, we require that

$$\begin{aligned} \mathrm{E}(\boldsymbol{\theta}'\hat{\boldsymbol{\beta}}) &= \mathrm{E}(\boldsymbol{\theta}'G\mathbf{y} + \boldsymbol{\theta}'\mathbf{d}) \\ &= \boldsymbol{\theta}'GX\boldsymbol{\beta} + \boldsymbol{\theta}'\mathbf{d} \\ &\equiv \boldsymbol{\theta}'\boldsymbol{\beta}. \end{aligned}$$

The identity in $\boldsymbol{\beta}$ is satisfied if and only if

$$\boldsymbol{\theta}' = \boldsymbol{\theta}'GX \qquad [6.5]$$

and $\boldsymbol{\theta}'\mathbf{d} = \mathbf{0}$ for all $\boldsymbol{\theta}$ such that $\boldsymbol{\theta}' = \boldsymbol{\alpha}'X$. The conditions for unbiasedness may therefore be expressed as $XGX = X$ and $X\mathbf{d} = \mathbf{0}$.

All unbiased estimates of $\boldsymbol{\theta}'\boldsymbol{\beta}$ are therefore of the form $\boldsymbol{\theta}'G\mathbf{y}$, where $G = X^{g_1}$. To obtain the conditions upon G for $\boldsymbol{\theta}'G\mathbf{y}$ to be a minimum variance estimate we use the following theorem due to Zyskind (1967). Since the result is also used in the next chapter, we prove it for the case where $\mathbf{y}$ has variance matrix V.

THEOREM 6.2 An unbiased estimate $\mathbf{w}'\mathbf{y} = \boldsymbol{\theta}'G\mathbf{y}$ of $\boldsymbol{\theta}'\boldsymbol{\beta}$ has minimum variance if and only if $V\mathbf{w}$ is contained in the column-space of X.

Proof: Let $\boldsymbol{\lambda}$ be a vector of Lagrange multipliers; then the minimum value of $\mathrm{var}(\mathbf{w}'\mathbf{y}) = \mathbf{w}'V\mathbf{w}$, subject to $\mathbf{w}'X = \boldsymbol{\theta}'GX = \boldsymbol{\theta}'$, is given by $\delta t/\delta\mathbf{w}' = 0$, where $t = \mathbf{w}'V\mathbf{w} - 2(\mathbf{w}'X - \boldsymbol{\theta}')\boldsymbol{\lambda}$. This gives the condition $V\mathbf{w} = X\boldsymbol{\lambda}$.

However, since the method of Lagrange multipliers gives only necessary conditions for a minimum, it must still be shown that $\mathbf{w}'V\mathbf{w}$

is actually a minimum under $V\mathbf{w} = X\boldsymbol{\lambda}$. Let $(\mathbf{w} + \mathbf{u})'\mathbf{y}$ be some other unbiased estimate of $\boldsymbol{\theta}'\boldsymbol{\beta}$. Since $\mathbf{w}'\mathbf{y}$ is also an unbiased estimate, this implies $\mathbf{u}'X = \mathbf{0}$. Thus

$$\begin{aligned}\operatorname{var}\{(\mathbf{w} + \mathbf{u})'\mathbf{y}\} &= \mathbf{w}'V\mathbf{w} + 2\mathbf{u}'(V\mathbf{w}) + \mathbf{u}'V\mathbf{u} \\ &= \mathbf{w}'V\mathbf{w} + 2\mathbf{u}'(X\boldsymbol{\lambda}) + \mathbf{u}'V\mathbf{u} \\ &= \mathbf{w}'V\mathbf{w} + \mathbf{u}'V\mathbf{u} \quad \text{since } \mathbf{u}'X = \mathbf{0},\end{aligned}$$

whence $\operatorname{var}(\mathbf{w}'\mathbf{y})$ is a minimum.

THEOREM 6.3 A BLU estimate of $\boldsymbol{\theta}'\boldsymbol{\beta}$ is given by $\boldsymbol{\theta}'G\mathbf{y}$ if and only if

$$\left.\begin{aligned} XGX &= X \\ \text{and} \quad (XG)' &= XG. \end{aligned}\right\} \qquad [6.6]$$

Proof: Theorem 6.1 established that $XGX = X$ is necessary and sufficient for $\boldsymbol{\theta}'G\mathbf{y}$ to be unbiased. Now, by Theorem 6.2 with $V = I$, $\boldsymbol{\theta}'G\mathbf{y}$ has minimum variance if and only if $G'\boldsymbol{\theta}$ is contained in the column-space of X, for all $\boldsymbol{\theta}'$ contained in the row-space of X, i.e. $\boldsymbol{\theta}'G\mathbf{y}$ has minimum variance if and only if $(XG)'$ is contained in the column-space of X. Now by Theorem 1.5, this condition is equivalent to

$$(XG)' = XG(XG)'$$

or

$$(I - XG)(XG)' = 0,$$

i.e. XG is symmetric.

Corollary 1 Any one of the following sets of conditions is necessary and sufficient for $\boldsymbol{\theta}'G\mathbf{y}$ to be a BLU estimate of an estimable linear function $\boldsymbol{\theta}'\boldsymbol{\beta}$:

(a) $XGX = X,\ (XG)' = XG$

(b) $XGX = X,\ (I - XG)(XG)' = 0$

(c) $X'XG = X'$

(d) $XG = XX^{g_3} = XX^{g}$. **[6.7]**

Proof: It has already been shown that (a) and (b) are equivalent, and obviously (a) implies (c), while (d) implies (a). It remains to demonstrate that (c) implies (d). If $X'XG = X'$, then by Theorem 1.6, Corollary 2, $G = (X'X)^{g_1}X' + [I - (X'X)^{g_1}X'X]Z$, whence $XG = X(X'X)^{g_1}X' = XX^{g_3} = XX^{g}$ by Theorem 2.13, Corollary 2.

Under the conditions of this corollary $\boldsymbol{\theta}'G\mathbf{y}$ is unique since by [6.5]

$$\begin{aligned} \boldsymbol{\theta}' G\mathbf{y} &= \boldsymbol{\theta}' GXG\mathbf{y} \\ &= \boldsymbol{\theta}' GXX^{g}\mathbf{y} \quad \text{by [6.7]} \\ &= \boldsymbol{\theta}' X^{g}\mathbf{y} \quad \text{by [6.3].} \end{aligned}$$

From the foregoing it is apparent that G is a new type of g-inverse satisfying conditions (1) and (3) only. However, the important feature of G seems to lie in relation [6.7], namely that XG is unique, symmetric idempotent.

Corollary 2 The BLU estimate of $\boldsymbol{\theta}'\boldsymbol{\beta}$ has variance $\sigma^2\boldsymbol{\theta}' S^{g_1}\boldsymbol{\theta}$, which is unique.

Proof:
$$\begin{aligned} \text{var}(\boldsymbol{\theta}' G\mathbf{y}) &= \sigma^2\boldsymbol{\theta}' GG'\boldsymbol{\theta} \\ &= \sigma^2\boldsymbol{\theta}' GXGG'X'G'\boldsymbol{\theta} \quad \text{by [6.5]} \\ &= \sigma^2\boldsymbol{\theta}' GXX^{g_3}(X^{g_3})'X'G'\boldsymbol{\theta} \quad \text{by [6.7]} \\ &= \sigma^2\boldsymbol{\theta}' X^{g_3}(X^{g_3})'\boldsymbol{\theta} \quad \text{by [6.5]} \\ &= \sigma^2\boldsymbol{\theta}' S^{g_1}S(S^{g_1})'\boldsymbol{\theta} \quad \text{by [2.10]} \\ &= \sigma^2\boldsymbol{\theta}' S^{g_1}\boldsymbol{\theta} \quad \text{by [6.4]} \end{aligned}$$

Alternatively $X^{g_3}(X^{g_3})' = S^{g_2}$ by the miscellaneous relationship (h) at the end of § 2.6; or again the substitution XX^{g} for XG instead of XX^{g_3}, gives the variance as $\sigma^2\boldsymbol{\theta}' S^{g}\boldsymbol{\theta}$. By Theorem 2.17, these expressions for the variance are the same, i.e. $\boldsymbol{\theta}' S^{g_1}\boldsymbol{\theta}$ is unique, since $\boldsymbol{\theta}'$ is contained in the row-space of S.

Corollary 3 The BLU estimate of $\boldsymbol{\theta}'\boldsymbol{\beta}$ is given by $\boldsymbol{\theta}'\mathbf{b}$, where $\mathbf{b}$ is any solution to the normal equations $S\mathbf{b} = X'\mathbf{y}$.

Proof: By Theorem 1.6, Corollary 4, any solution to the normal equations is given by

$$\mathbf{b} = S^{g_1}X'\mathbf{y} + (I - S^{g_1}S)\mathbf{z}.$$

By [6.4] $\boldsymbol{\theta}'\mathbf{b} = \boldsymbol{\theta}' S^{g_1}X'\mathbf{y} = \boldsymbol{\theta}' X^{g_3}\mathbf{y}$, so that [6.7] is satisfied.

This verifies that the choice $G = X^{g_3}$ is permissible in terms of [6.6], but it is not necessary for G to satisfy condition (2) in addition to conditions (1) and (3). However, if G is of the form BX', condition (2) is also satisfied since $GXG = BX'\underline{XBX'} = B\underline{X'XB'X'} = BX' = G$.

Nor, in fact, is $G = X^{g_3}$ necessary for $G\mathbf{y}$ to be a solution to the normal equations. This is shown by the following result (cf. Rao, 1967).

THEOREM 6.4 Let $\mathbf{b} = G\mathbf{y} + \mathbf{d}$. Then $\mathbf{b}$ is a least-squares solution of the observational equations $X\mathbf{b} = \mathbf{y}$ if and only if G satisfies any of the conditions of Theorem 6.3, Corollary 1, and $X\mathbf{d} = \mathbf{0}$.

Proof: Application of the least-squares criterion by finding $\mathbf{b}$ so as to minimize $(\mathbf{y} - X\mathbf{b})'(\mathbf{y} - X\mathbf{b})$ leads to the normal equations $S\mathbf{b} = X'\mathbf{y}$, i.e. $\mathbf{b}$ is a least-squares solution if and only if it satisfies $S\mathbf{b} = X'\mathbf{y}$, i.e. if and only if $SG\mathbf{y} + S\mathbf{d} = X'\mathbf{y}$ for all $\mathbf{y}$, i.e. if and only if $SG = X'$ and $S\mathbf{d} = \mathbf{0}$. The former is (c) of Theorem 6.3, Corollary 1, and the latter is equivalent to $X\mathbf{d} = \mathbf{0}$.

In effect, then, although from Theorem 6.3, Corollary 3, $\mathbf{b} = X^{g_3}\mathbf{y} + (I - X^{g_3}X)\mathbf{z}$ is a general solution to the normal equations, the general solution may be "more basically" represented by

$$\mathbf{b} = G\mathbf{y} + (I - X^{g_1}X)\mathbf{z}, \qquad [6.8]$$

where G is a g_1-inverse of X satisfying conditions (1) and (3).

Corollary The fitted values, $\hat{\mathbf{y}} = X\mathbf{b}$, are unique.

Proof: By [6.8] $\hat{\mathbf{y}} = X\mathbf{b} = XG\mathbf{y} = XX^{g}\mathbf{y}$ (by [6.7]), which is unique.

From Theorems 6.3 and 6.4 a "Gauss–Markoff" theorem for model [6.1] is now apparent:

THEOREM 6.5 The BLU estimate of an estimable linear function $\boldsymbol{\theta}'\boldsymbol{\beta}$ is given by $\boldsymbol{\theta}'\mathbf{b}$ if and only if $\mathbf{b}$ is a least-squares solution of the observational equations $X\mathbf{b} = \mathbf{y}$.

If the data to which the model [6.1] is to be fitted are obtained from a well-designed experiment, there is little problem in determining which functions of $\boldsymbol{\beta}$ are estimable, since the matrix X will take some well-known predetermined form. However, if the data are irregular – for example, if there are several missing observations – then [6.4] could constitute a direct test of whether some linear function of interest is estimable.

Searle (1966, pp. 262–5) points out that all linear functions of the form $\boldsymbol{\alpha}'S^{g_1}S\boldsymbol{\beta}$, where $\boldsymbol{\alpha}$ is an arbitrary k-dimensional vector, are estimable. This follows from [6.4] since $\boldsymbol{\alpha}'S^{g_1}SS^{g_1}S = \boldsymbol{\alpha}'S^{g_1}S$. Thus, by assigning specific values to the elements of $\boldsymbol{\alpha}$, one could generate specific estimable functions. Furthermore, Searle suggests that in order to determine whether a linear function $\boldsymbol{\theta}'\boldsymbol{\beta}$ is estimable, one would see if values of $\boldsymbol{\alpha}$ could be found, such that $\boldsymbol{\alpha}'S^{g_1}S = \boldsymbol{\theta}'$. However, both these proposals are unnecessarily complicated. All linear functions of the form $\mathbf{a}'X\boldsymbol{\beta}$ or $\mathbf{a}'S\boldsymbol{\beta}$ are estimable, and thus either of these could serve as "generators" of estimable functions. To test whether a specific linear function of interest is estimable, one would simply apply the condition [6.4].

The invariance of the BLU estimate $\boldsymbol{\theta}'G\mathbf{y}$ means that in numerical applications G may be taken as any g-inverse of X satisfying at least conditions (1) and (3). It would appear that if S is partitioned as

$$S = \begin{bmatrix} S_{11} & S_{12} \\ S_{21} & S_{22} \end{bmatrix},$$

where S_{11} is $p \times p$ of rank p, then the selection of $G = S^{g_2}X' = X^{g_3}$, with

$$S^{g_2} = \begin{bmatrix} S_{11}^{-1} & 0 \\ 0 & 0 \end{bmatrix}, \qquad [6.9]$$

will usually lead to the least arithmetic (cf. [4.15] and the discussion following).

If $\boldsymbol{\theta}$ and X are partitioned to conform with the above partitioning of S, then the substitution of [6.9] in [6.4] shows that $\boldsymbol{\theta}'\boldsymbol{\beta}$ is estimable if and only if $\boldsymbol{\theta}_2' = \boldsymbol{\theta}_1' S_{11}^{-1} S_{12}$. Similarly, from Theorem 6.3, Corollary 3, the BLU estimate of $\boldsymbol{\theta}'\boldsymbol{\beta}$ is $\boldsymbol{\theta}_1' S_{11}^{-1} X_1' \mathbf{y}$. These expressions for the estimability condition and the BLU estimate of an estimable linear function were derived by Roy (1953), who also showed that they do not depend on the choice of a basis for X. This is, of course, an alternative interpretation of the invariance under choice of g-inverse.

Before discussing tests of hypotheses we derive some results on the regression and error sums of squares.

THEOREM 6.6

(a) The sum of squares due to fitting model [6.1], the "regression" sum of squares, is given by $\mathbf{b}'S\mathbf{b}$, where $\mathbf{b}$ is any solution to $S\mathbf{b} = X'\mathbf{y} = \mathbf{g}$.

(b) The residual, or "error", sum of squares may be expressed as

$$SS(E) = \mathbf{y}'(I - XS^{g_1}X')\mathbf{y},$$

and is unique.

(c) $SS(E)$ has $\chi^2\sigma^2$ distribution with $n - p$ degrees of freedom if $\mathbf{y}$ is $\mathrm{N}(X\boldsymbol{\beta}, \sigma^2 I)$.

Proof: By Theorem 6.4, Corollary, the residual sum of squares is given by $\mathbf{y}'(I - XG)'(I - XG)\mathbf{y} = \mathbf{y}'(I - XG)\mathbf{y} = \mathbf{y}'(I - XX^{g_3})\mathbf{y} = \mathbf{y}'(I - XS^{g_1}X')\mathbf{y}$, and is unique.

The regression sum of squares is given by $\mathbf{y}'XG\mathbf{y} = \mathbf{b}'\mathbf{g}$ and also by $\mathbf{y}'XS^{g_1}X'\mathbf{y} = \mathbf{g}'S^{g_1}\mathbf{g}$. Since XG is symmetric idempotent, this sum of squares is also given by $\mathbf{y}'(XG)'XG\mathbf{y} = \mathbf{y}'G'SG\mathbf{y} = \mathbf{b}'S\mathbf{b}$.

As regards the distribution of $SS(E)$, since $I - XG$ is symmetric idempotent and $(I - XG)X\boldsymbol{\beta} = \mathbf{0}$, it follows from Theorem 5.3, Corollary 1, (with $V = I$, $\mathbf{m} = \mathbf{0}$, and $d = 0$) that $SS(E)$ has $\chi^2\sigma^2$ distribution with degrees of freedom given by $\mathrm{r}(I - XG) = n - p$ by Theorem 2.1(a).

Tests of hypotheses

The test of the hypothesis $\boldsymbol{\beta} = \mathbf{0}$ based on the partitioning of the total sum of squares into regression and error components (Theorem 6.6) is well known and will not be discussed.

We consider here the test of the general linear hypothesis $L\boldsymbol{\beta} = \mathbf{z}$, *where L is contained in the row-space of S*, i.e. the linear functions $L\boldsymbol{\beta}$ are estimable and the hypothesis is therefore testable. The vector $\mathbf{z}$ consists of known constants and is contained in the column-space of L, i.e. the hypothesis is consistent. Usually in analysis of variance $\mathbf{z} = \mathbf{0}$.

Searle (1965; also 1966, pp. 275–9) and Rao (1965, pp. 155–7) have derived the sum of squares due to the hypothesis when $\mathrm{r}(L) = q \leqslant k$. However, no restriction upon the rank of L will be imposed here, i.e. L will be considered to be a $q \times k$ matrix of rank $s \leqslant q \leqslant k$.

Following Searle and Rao, the sum of squares due to the hypothesis, which will be denoted by $SS(H)$, is obtained as

$$SS(H) = SS(E^{\dagger}) - SS(E),$$

where $SS(E^{\dagger})$ denotes the residual sum of squares after fitting model [6.1] amended by the hypothesis, and $SS(E)$ represents the residual sum of squares after fitting model [6.1] unamended. Fitting the amended model is equivalent to fitting the full model [6.1], subject to the restriction $L\boldsymbol{\beta} = \mathbf{z}$. The constrained sum of squared residuals is obtained by applying the Lagrangian technique of undetermined multipliers, which leads to equations

$$\begin{bmatrix} S & L' \\ L & 0 \end{bmatrix} \begin{bmatrix} \mathbf{b}^{\dagger} \\ \boldsymbol{\lambda} \end{bmatrix} = \begin{bmatrix} \mathbf{g} \\ \mathbf{z} \end{bmatrix}, \qquad [6.10]$$

where $S = X'X$, $\mathbf{g} = X'\mathbf{y}$, and $\boldsymbol{\lambda}$ is a vector of Lagrange multipliers.

Since L is contained in the row-space of S, the expression for a g_1-inverse of a bordered matrix given in [3.44] may be used to solve [6.10]. Thus

$$\mathbf{b}^{\dagger} = \mathbf{b} - S^{g_1}L'(LS^{g_1}L')^{g_1}(L\mathbf{b} - \mathbf{z}),$$

and

$$\boldsymbol{\lambda} = (LS^{g_1}L')^{g_1}(L\mathbf{b} - \mathbf{z}),$$

where $\mathbf{b} = S^{g_1}\mathbf{g}$.

Now $SS(E^\dagger) = (\mathbf{y} - X\mathbf{b}^\dagger)'(\mathbf{y} - X\mathbf{b}^\dagger)$, and substitution for $\mathbf{b}^\dagger$ gives, with a little simplification,

$$SS(E^\dagger) = (\mathbf{y} - X\mathbf{b})'(\mathbf{y} - X\mathbf{b}) + (L\mathbf{b} - \mathbf{z})'(Q^{g_1})'QQ^{g_1}(L\mathbf{b} - \mathbf{z}),$$

where a further substitution $Q = LS^{g_1}L'$ has been made. The first term in $SS(E^\dagger)$ is, of course, $SS(E)$, whence

$$SS(H) = SS(E^\dagger) - SS(E) = (L\mathbf{b} - \mathbf{z})'Q^*(L\mathbf{b} - \mathbf{z}), \qquad [6.11]$$

where $Q^* = (Q^{g_1})'QQ^{g_1}$ is a symmetric, positive semidefinite g_2-inverse of Q.

Before deriving the distributional properties of this sum of squares, it will be shown that it is unique. It was shown in Theorem 6.6 (b) that $SS(E)$ is unique, so that [6.11] is unique if $SS(E^\dagger)$ is invariant under the choice of solution of [6.10], i.e. we must show that $X\mathbf{b}^\dagger$ is unique.

If $M = \begin{bmatrix} S & L' \\ L & 0 \end{bmatrix}$, then by Theorem 1.6, Corollary 5, $X\mathbf{b}^\dagger = [X \quad 0]\begin{bmatrix} \mathbf{b}^\dagger \\ \boldsymbol{\lambda} \end{bmatrix}$

is unique if and only if $[X \quad 0]M^{g_1}M = [X \quad 0]$. It is easily verified from [3.44] that

$$M^{g_1}M = \begin{bmatrix} S^{g_1}S & 0 \\ 0 & Q^{g_1}Q \end{bmatrix},$$

whereupon the result follows.

It will now be established that the ratio

$$\frac{SS(H)}{n_1}\bigg/\frac{SS(E)}{n_2},$$

where n_1 and n_2 are appropriate divisors, follows the F-distribution. It is assumed, of course, that $\mathbf{y}$ is $N(X\boldsymbol{\beta}, \sigma^2 I)$.

The substitution $\mathbf{b} = S^{g_1}X'\mathbf{y}$ in [6.11] gives

$$\begin{aligned} SS(H) &= \mathbf{y}'X(S^{g_1})'L'Q^*LS^{g_1}X'\mathbf{y} - 2\mathbf{z}'Q^*LS^{g_1}X'\mathbf{y} + \mathbf{z}'Q^*\mathbf{z} \\ &= \mathbf{y}'W\mathbf{y} + \mathbf{t}'\mathbf{y} + d \quad \text{(say).} \end{aligned}$$

Then

$$\begin{aligned} W^2 &= X(S^{g_1})'L'Q^*LS^{g_1}S(S^{g_1})'L'Q^*LS^{g_1}X' \\ &= X(S^{g_1})'L'Q^*QQ^*LS^{g_1}X' \quad (\text{since } LS^{g_1}S = L) \\ &= X(S^{g_1})'L'Q^*LS^{g_1}X' \quad (\text{since } Q^* = Q^{g_2}) \\ &= W; \end{aligned}$$

similarly,

$$\mathbf{t}'W = -2\mathbf{z}'Q^*LS^{g_1}S(S^{g_1})'L'Q^*LS^{g_1}X'$$
$$= -2\mathbf{z}'Q^*LS^{g_1}X'$$
$$= \mathbf{t}',$$

and

$$\tfrac{1}{4}\mathbf{t}'\mathbf{t} = \mathbf{z}'Q^*LS^{g_1}S(S^{g_1})'L'Q^*\mathbf{z}$$
$$= \mathbf{z}'Q^*\mathbf{z} = d.$$

Thus by Theorem 5.2 $SS(H)$ has noncentral $\chi^2\sigma^2$ distribution. The degrees of freedom are

$$\operatorname{tr}(W) = \operatorname{tr}[S(S^{g_1})'L'Q^*LS^{g_1}]$$
$$= \operatorname{tr}(Q^*LS^{g_1}L')$$
$$= \operatorname{tr}(Q^*Q)$$
$$= \operatorname{r}(Q) \quad \text{(by Theorem 2.1)},$$
$$= s \quad \text{(by Theorem 2.17)},$$

and the noncentrality parameter is

$$(\sigma^2)^{-1}\{\boldsymbol{\beta}'X'[X(S^{g_1})'L'Q^*LS^{g_1}X']X\boldsymbol{\beta} - 2\mathbf{z}'(Q^*LS^{g_1}X')X\boldsymbol{\beta} + \mathbf{z}'Q^*\mathbf{z}\}$$
$$= (\sigma^2)^{-1}\{\boldsymbol{\beta}'L'Q^*L\boldsymbol{\beta} - 2\mathbf{z}'Q^*L\boldsymbol{\beta} + \mathbf{z}'Q^*\mathbf{z}\},$$

which is zero under the hypothesis $L\boldsymbol{\beta} = \mathbf{z}$.

Finally, since by [2.6]

$$W(I - XS^{g_1}X') = 0$$

and

$$\mathbf{t}'(I - XS^{g_1}X') = \mathbf{0},$$

the distributions of $SS(H)$ and $SS(E)$ are independent (Laha, 1956). Thus, by Theorem 6.6 (c), the ratio

$$\frac{SS(H)}{s}\bigg/\frac{SS(E)}{n-p}$$

follows the F-distribution with s and $n - p$ degrees of freedom. This ratio reduces to that obtained by Searle if Q^{g_1} is replaced by Q^{-1} and $s = \operatorname{r}(L)$ is replaced by q. Searle also unnecessarily restricted S^{g_1} to being symmetric.

Reparametrization

If the vector of parameters is transformed by $\boldsymbol{\beta}^* = T\boldsymbol{\beta}$, thereby inducing a contragredient transformation of X to $X^* = XU$ such that $X^*\boldsymbol{\beta}^* = X\boldsymbol{\beta}$, this constitutes a reparametrization of the model.

To achieve a reparametrization for a given T, U must be found to satisfy

$$XUT = X. \qquad [6.12]$$

For consistency we must have by [1.12]

$$XX^{g_1}XT^{g_1}T = X,$$

or

$$XT^{g_1}T = X, \qquad [6.13]$$

i.e. the row-space of T must contain that of X (Theorem 1.5, Corollary). If T satisfies this condition,

$$U = X^{g_1}XT^{g_1} + Z - X^{g_1}XZTT^{g_1}$$

by [1.13] and

$$X^* = X\{T^{g_1} + Z(I - TT^{g_1})\}.$$

Since $X^*T = X$ and $X^* = XU$, $\mathrm{r}(X^*) = \mathrm{r}(X) = p$. Thus under this definition the reparametrized model retains its rank, but a model of reduced rank is obtainable by equating parameters to zero, as covered in § 6.6.

The reparametrized model is equivalent in every way to the original model. Any estimable function in terms of the reparametrized model is given by $\mathbf{a}'X^*\boldsymbol{\beta}^* = \mathbf{a}'X\boldsymbol{\beta}$, and is therefore estimable in terms of the original model. Its BLU estimate may therefore be written as $\mathbf{a}'X\mathbf{b}$, where $\mathbf{b}$ is any solution to $S\mathbf{b} = \mathbf{g}$. The normal equations for the reparametrized model are

$$S^*\mathbf{b}^* = \mathbf{g}^*,$$

i.e.

$$U'SU\mathbf{b}^* = U'\mathbf{g};$$

that $\mathbf{b}^* = T\mathbf{b}$ is a solution is seen from

$$\begin{aligned} U'SUT\mathbf{b} &= U'S\mathbf{b} \quad \text{by [6.12]} \\ &= U'\mathbf{g}. \end{aligned}$$

Hence the BLU estimate of $\mathbf{a}'X^*\boldsymbol{\beta}^*$ is $\mathbf{a}'X^*T\mathbf{b} = \mathbf{a}'X\mathbf{b}$. The variance of $\mathbf{a}'X^*\mathbf{b}^*$ is $\sigma^2\mathbf{a}'XS^{g_1}X'\mathbf{a} = \mathbf{a}'X^*TS^{g_1}T'(X^*)'\mathbf{a}$, and since

$$\begin{aligned} S^*(TS^{g_1}T')S^* &= U'SUTS^{g_1}T'U'SU \\ &= U'SS^{g_1}SU \quad \text{by [6.12]} \\ &= S^*, \end{aligned}$$

it follows that $\mathrm{var}(\mathbf{a}'X^*\mathbf{b}^*) = \mathbf{a}'X^*(S^*)^{g_1}(X^*)'\mathbf{a}$. Also the sum of squares due to regression is

$$\begin{aligned} (\mathbf{b}^*)'S^*\mathbf{b}^* &= \mathbf{b}'T'U'SUT\mathbf{b} \\ &= \mathbf{b}'S\mathbf{b}. \end{aligned}$$

If T is chosen so that $\boldsymbol{\beta}^*$ is estimable, $TX^{g_1}X = T$ by [6.3], and from [6.13] $\mathrm{r}(T) = \mathrm{r}(X) = p$. In this case any linear function of $\boldsymbol{\beta}^*$ is estimable. The BLU estimate of $\mathbf{a}'\boldsymbol{\beta}^*$ is $\mathbf{a}'\mathbf{b}^*$ with variance $\sigma^2\mathbf{a}'TS^{g_1}T'\mathbf{a} = \sigma^2\mathbf{a}'TS^{g}T'\mathbf{a}$ (by Theorem 2.17); it may be verified that $TS^{g}T'$ is a symmetric g_2-inverse of S^*.

Graybill (1961, pp. 235–9) discusses the full-rank reparametrization in which $\boldsymbol{\beta}^*$ consists of p linearly independent estimable functions of $\boldsymbol{\beta}$. For this case:

T is $p \times k$ of full row-rank; $TT^{g_1} = I$;
$X^* = XT^{g_1}$ is $n \times p$ of full column-rank; $(X^*)^{g_1}X^* = I$;
$TU = I$, i.e. U has full column-rank and $U = T^{g_1}$;
S^* is nonsingular.

Graybill defines U as Z_1 in

$$Z = [Z_1 \quad Z_2] = \begin{bmatrix} T \\ L \end{bmatrix}^{-1}$$

where L is complementary to T and of full row-rank, i.e.

$$U = T^{g_1} = (T'T + L'L)^{-1}T'$$

by Theorem 2.19 (c). Further, X^* is unique, since, if G_1 and G_2 are any two g_1-inverses of T, $XG_1T = X$, and postmultiplication by G_2 gives $XG_1TG_2 = XG_1 = XG_2$.

6.3 The method of imposed linear restrictions

In the second approach to the estimation problem for the linear model of less than full rank, an estimator $\hat{\boldsymbol{\beta}}^*$ is obtained which is the minimum variance, linear, conditionally unbiased estimate of $\boldsymbol{\beta}$ subject to $L\boldsymbol{\beta} = \mathbf{c}$, *where L is a $q \times k$ matrix complementary to X*, and $\mathbf{c}$ is in the column-space of L, i.e. the restrictions are consistent.

The principle of least squares may be applied by minimizing the residual sum of squares $(\mathbf{y} - X\hat{\boldsymbol{\beta}}^*)'(\mathbf{y} - X\hat{\boldsymbol{\beta}}^*)$, subject to $L\hat{\boldsymbol{\beta}}^* = \mathbf{c}$. Introduction of a vector of Lagrange multipliers $\boldsymbol{\lambda}$ leads to the equations

$$\begin{bmatrix} S & L' \\ L & 0 \end{bmatrix}\begin{bmatrix} \hat{\boldsymbol{\beta}}^* \\ \boldsymbol{\lambda} \end{bmatrix} = \begin{bmatrix} \mathbf{g} \\ \mathbf{c} \end{bmatrix} \qquad [6.14]$$

where $S = X'X$ and $\mathbf{g} = X'\mathbf{y}$. Thus the procedure for forming the normal equations ignoring the restrictions may be followed, and a bordering process applied to give the matrix

$$M = \begin{bmatrix} S & L' \\ L & 0 \end{bmatrix}. \qquad [6.15]$$

The general solution to [6.14] is given by

$$\begin{bmatrix} \hat{\boldsymbol{\beta}}^* \\ \boldsymbol{\lambda} \end{bmatrix} = M^{g_1}\begin{bmatrix} \mathbf{g} \\ \mathbf{c} \end{bmatrix} + (I - M^{g_1}M)\begin{bmatrix} \mathbf{z}_1 \\ \mathbf{z}_2 \end{bmatrix},$$

where M^{g_1} is any g_1-inverse of M, and for convenience we take M^{g_1} as the g-inverse of M given by [3.47], in which $K^{-1} = (S + L'L)^{-1}$. It is readily verified using the results of Theorem 2.19 (a) and (b), namely $XK^{-1}L' = 0$ and $SK^{-1}S = S$, that

$$M^g M = \begin{bmatrix} I & 0 \\ 0 & LK^{-1}L' \end{bmatrix},$$

whence

$$\hat{\boldsymbol{\beta}}^* = K^{-1}SK^{-1}X'\mathbf{y} + K^{-1}L'\mathbf{c},$$

which does not involve $\mathbf{z}$ and is therefore unique, and

$$\begin{aligned} \boldsymbol{\lambda} &= LK^{-1}X'\mathbf{y} + (I - LK^{-1}L')\mathbf{z}_2 \\ &= (I - LK^{-1}L')\mathbf{z}_2 . \end{aligned}$$

By [2.6],

$$\hat{\boldsymbol{\beta}}^* = K^{-1}X'\mathbf{y} + K^{-1}L'\mathbf{c} \qquad [6.16]$$

$$= X^{\ddagger}\mathbf{y} + L^{\ddagger}\mathbf{c}, \qquad [6.17]$$

where $X^{\ddagger} = K^{-1}X' = X^{g_3}$ and $L^{\ddagger} = K^{-1}L' = L^{g_3}$.

The expression [6.17] was obtained (for the case $r(L) = q$) by Plackett (1950), while the portrayal of Plackett's solution in terms of g_3-inverses of X and L was presented by Chipman (1964), who considered the case var$(\mathbf{y}) = V$, positive definite.

If L is chosen to be an orthogonal complement of X, then by Theorem 2.19, Corollary, [6.16] becomes

$$\hat{\boldsymbol{\beta}}^* = X^g\mathbf{y} + L^g\mathbf{c}.$$

The following theorems establish that $\hat{\boldsymbol{\beta}}^*$ has the required properties.

THEOREM 6.7 A set of necessary and sufficient conditions for

$G\mathbf{y} + \mathbf{d}$ to be a conditionally unbiased estimate of $\boldsymbol{\beta}$ subject to $L\boldsymbol{\beta} = \mathbf{c}$ is given by

$$\left.\begin{aligned} GX &= I - \Lambda L \\ \Lambda\mathbf{c} &= \mathbf{d}, \end{aligned}\right\} \qquad [6.18]$$

where L is complementary to X and Λ is a matrix of Lagrange multipliers (Chipman, 1964).

Proof: The criterion of conditional unbiasedness requires

$$\begin{aligned} \mathrm{E}(G\mathbf{y} + \mathbf{d}) &= GX\boldsymbol{\beta} + \mathbf{d} \\ &\equiv \boldsymbol{\beta} + \Lambda(\mathbf{c} - L\boldsymbol{\beta}) \\ &\equiv (I - \Lambda L)\boldsymbol{\beta} + \Lambda\mathbf{c}, \end{aligned}$$

and the identity in $\boldsymbol{\beta}$ holds if and only if [6.18] holds, the equation $GX + \Lambda L = I$ being possible only if L is complementary to X.

THEOREM 6.8 The minimum variance, linear, conditionally unbiased estimate of $\boldsymbol{\beta}$ subject to $L\boldsymbol{\beta} = \mathbf{c}$ is $\hat{\boldsymbol{\beta}}^* = X^{\ddagger}\mathbf{y} + L^{\ddagger}\mathbf{c}$.

Proof: By Theorem 6.7 it follows from [2.14] with $G = X^{\ddagger}$ and $\Lambda = L^{\ddagger}$ that [6.18] is satisfied, i.e. $\hat{\boldsymbol{\beta}}^*$ is conditionally unbiased.

Now consider any other estimator $G\mathbf{y} + \mathbf{d}$ such that $GX = I - \Lambda L$. Then by Theorem 2.19 (a)

$$GXX^{\ddagger} = X^{\ddagger}$$

and

$$\begin{aligned} \operatorname{var}(G\mathbf{y} + \mathbf{d}) = \sigma^2 GG' &= \sigma^2\{X^{\ddagger} + (G - X^{\ddagger})\}\{X^{\ddagger} + (G - X^{\ddagger})\}' \\ &= \sigma^2\{X^{\ddagger}(X^{\ddagger})' + (G - X^{\ddagger})(G - X^{\ddagger})'\}, \end{aligned}$$

whence it follows that the variance is minimal when $G = X^{\ddagger}$.

In analysis of variance it is usual for the number of redundancies and the number of restrictions to be in balance, i.e. $\mathrm{r}(L) = q = k - p$. For this particular case $\boldsymbol{\lambda} = (I - LK^{-1}L')\mathbf{z}_2 = (I - LL^{g_3})\mathbf{z}_2 = \mathbf{0}$, since L^{g_3} is a right inverse of L. The equations [6.14] therefore reduce to

$$\begin{bmatrix} S \\ L \end{bmatrix}\hat{\boldsymbol{\beta}}^* = \begin{bmatrix} \mathbf{g} \\ \mathbf{c} \end{bmatrix}. \qquad [6.19]$$

Premultiplication of both sides by $[I \quad L']$ gives

$$(S + L'L)\hat{\boldsymbol{\beta}}^* = \mathbf{g} + L'\mathbf{c},$$

whence $\hat{\boldsymbol{\beta}}^* = X^{\ddagger}\mathbf{y} + L^{\ddagger}\mathbf{c}$ as before. It is also usual in analysis of variance that $\mathbf{c} = \mathbf{0}$. The solution to equations [6.19] for this case was derived by Plackett (1960).

6.4 Relationship between estimability and linear restrictions

From [6.17],

$$S\hat{\boldsymbol{\beta}}^* = SK^{-1}X'\mathbf{y} + SK^{-1}L'\mathbf{c} = X'\mathbf{y},$$

i.e. $\hat{\boldsymbol{\beta}}^*$ is a solution to the normal equations. The linear restrictions may therefore be regarded as merely a device for obtaining a solution to the normal equations when S is singular. This viewpoint is sustained by the fact that L and $\mathbf{c}$ may be chosen at will, subject to the conditions given at the start of §6.3. This has led Searle (1967) to offer the following criticism of the method of imposed linear restrictions :-

> "Since it is sometimes not clear whether 'constraints' must apply to the model or whether they are merely a convenient artifact for obtaining a solution, the use of constraints can lead to confusion in understanding their meaning, their effect on the solution, and the consequence of being able to have many solutions."

Kempthorne (1952) illustrates both these forms of constraints, deriving a constrained model as a reparametrization of an original unconstrained model (p. 69). The simplest possible illustration of this is provided by the model for two independent random samples from infinite populations with different means. The primitive model may be represented as

$$\begin{bmatrix} \mathbf{y}_1 \\ \mathbf{y}_2 \end{bmatrix} = \begin{bmatrix} \mathbf{1}_1 & \mathbf{0} \\ \mathbf{0} & \mathbf{1}_2 \end{bmatrix} \begin{bmatrix} \mu_1 \\ \mu_2 \end{bmatrix} + \boldsymbol{\varepsilon},$$

where $\mathbf{y}_1$, $\mathbf{y}_2$ are the sample vectors from the two populations with means μ_1 and μ_2, and $\mathbf{1}_1$ and $\mathbf{1}_2$ are vectors with n_1 and n_2 elements, respectively, all unity. If μ_1 and μ_2 are replaced by $\frac{1}{2}(\mu_1 + \mu_2) - \frac{1}{2}(\mu_2 - \mu_1)$ and $\frac{1}{2}(\mu_1 + \mu_2) + \frac{1}{2}(\mu_2 - \mu_1)$, the familiar model

$$\left.\begin{aligned} \begin{bmatrix} \mathbf{y}_1 \\ \mathbf{y}_2 \end{bmatrix} &= \begin{bmatrix} \mathbf{1}_1 & \mathbf{1}_1 & \mathbf{0} \\ \mathbf{1}_2 & \mathbf{0} & \mathbf{1}_2 \end{bmatrix} \begin{bmatrix} \mu \\ \tau_1 \\ \tau_2 \end{bmatrix} + \boldsymbol{\varepsilon} \\ (\tau_1 + \tau_2 &= 0) \end{aligned}\right\} \qquad [6.20]$$

is obtained for which the normal equations are

$$\begin{bmatrix} n & n_1 & n_2 \\ n_1 & n_1 & 0 \\ n_2 & 0 & n_2 \end{bmatrix} \begin{bmatrix} \hat{\mu} \\ \hat{\tau}_1 \\ \hat{\tau}_2 \end{bmatrix} = \begin{bmatrix} \Sigma y \\ \Sigma y_1 \\ \Sigma y_2 \end{bmatrix}, \qquad [6.21]$$

where $n = n_1 + n_2$, and Σy is the total of the variate-values in both samples. While it would seem that the analysis of such restricted models should properly fall under the discussion of § 6.5 in that the parameters themselves are subject to constraints, the concept of reparametrization clearly envisages an original unrestricted model.

Yates (1933, 1934), who did not distinguish between parameters and their estimates or exhibit a formal model, used such natural restrictions to reduce the number of "constants" to be "fitted" to a minimum number of independent constants. This would be illustrated by taking $\tau_1 = -\tau_2$ or $-\tau_1 = \tau_2 = \tau$ in [6.20]. However, the type of reparametrization illustrated above and the relation of the restrictions back to the parameters of the model seem to be no longer fashionable, except perhaps in models for the treatment effects of 2^n factorial designs in terms of main effects and interactions. Hence the word "imposed" in the title of § 6.3 seems justified.

Yates's elimination of redundant constants is equivalent to solving [6.19] by using $L\hat{\boldsymbol{\beta}}^* = \mathbf{c}$ to eliminate q unknowns in $S\hat{\boldsymbol{\beta}}^* = \mathbf{g}$, thereby reducing the normal equations to a full-rank set. Quenouille (1950) presented a justification of this, but also advised that the restrictions did not need to be the natural ones, but could be chosen to best advantage to reduce the calculations. For example, if in [6.21] $n_1 = n_2$, the restriction $\hat{\tau}_1 + \hat{\tau}_2 = 0$ immediately produces the solution $\hat{\mu} = \bar{y}$, $\hat{\tau}_1 = \bar{y}_1 - \bar{y}$, $\hat{\tau}_2 = \bar{y}_2 - \bar{y}$; however, if $n_1 \neq n_2$, these restrictions do not give a quick solution, whereas $n_1\hat{\tau}_1 + n_2\hat{\tau}_2 = 0$ does. Quenouille substituted in $S\hat{\boldsymbol{\beta}}^* = \mathbf{g}$ from $L\hat{\boldsymbol{\beta}}^* = \mathbf{c}$, but without necessarily eliminating any unknowns. Graybill (1961, p. 248) discusses the solution of the normal equations by means of [6.19] and the choice of L to make the solution easy.

In the days when the equations were solved on desk calculating machines by iteration, these methods sufficed. The tediousness of iterative calculations when they converged slowly led to a preference for closed methods of solution such as the Doolittle or Cholesky, especially when these were shown to have other important advantages (e.g. Fox, 1950; Fox and Hayes, 1951; Rushton, 1951). With the closed methods the retention of symmetry has great advantages, and this can be achieved by solving [6.14], in practice with L of full row-rank as in [6.19] (e.g. Kempthorne 1952, p. 80).

However the use of [6.14] requires the solution of equations of increased order. Later it was realized that [6.19] could still be used and symmetry retained by taking the constraints in the form $\beta_j = 0$, where β_j represents $k - p$ redundant parameters. This amounts to simply dropping these parameters from the model (striking out corresponding rows and columns of S and corresponding elements of $\mathbf{g}$), which thereby becomes one of full rank.

The justification for solving [6.14] or [6.19] and the use of any convenient L complementary to X is thus seen to lie in Theorem 6.3, Corollary 3, since $\boldsymbol{\theta}'\hat{\boldsymbol{\beta}}^*$ obtained in this way is a BLU estimate of any estimable function $\boldsymbol{\theta}'\boldsymbol{\beta}$.

The same result may be demonstrated from the minimum variance conditionally unbiased viewpoint by the following theorem and its corollary:

THEOREM 6.9 A linear estimator $\tilde{\boldsymbol{\beta}} = G\mathbf{y} + \mathbf{d}$ provides an unbiased estimator $\boldsymbol{\theta}'\tilde{\boldsymbol{\beta}}$ of any estimable function $\boldsymbol{\theta}'\boldsymbol{\beta}$ if and only if $\tilde{\boldsymbol{\beta}}$ is a conditionally unbiased estimator of $\boldsymbol{\beta}$, subject to $L\boldsymbol{\beta} = \mathbf{c}$, where L is some matrix complementary to X and $\mathbf{c}$ is a vector in the column-space L (Chipman, 1964).

Proof: By Theorem 6.1, $\boldsymbol{\theta}'\tilde{\boldsymbol{\beta}} = \boldsymbol{\theta}'(G\mathbf{y} + \mathbf{d})$ is an unbiased estimate of $\boldsymbol{\theta}'\boldsymbol{\beta}$ if and only if

$$XGX = X \quad \text{and} \quad X\mathbf{d} = \mathbf{0}.$$

On the other hand, the necessary and sufficient conditions for $\tilde{\boldsymbol{\beta}} = G\mathbf{y} + \mathbf{d}$ to be a conditionally unbiased estimator of $\boldsymbol{\beta}$ subject to $L\boldsymbol{\beta} = \mathbf{c}$ are as given by [6.18], namely

$$GX = I - \Lambda L \quad \text{and} \quad \Lambda\mathbf{c} = \mathbf{d}.$$

Now by Theorem 2.20, $G = X^{g_1}$ if and only if $GX = K^{-1}S = X^{\ddagger}X$, and by [2.14] $X^{\ddagger}X = I - L^{\ddagger}L$. Thus the conditions on G are equivalent.

Moreover, if $G = X^{g_1}$ and $X\mathbf{d} = \mathbf{0}$, then by Theorem 1.6, Corollary 4, $\mathbf{d} = (I - GX)\mathbf{z}$, where $\mathbf{z}$ is arbitrary. Since $I - GX = \Lambda L$, $\mathbf{d} = \Lambda L\mathbf{z} = \Lambda\mathbf{c}$. Conversely, if $\mathbf{d} = \Lambda\mathbf{c}$, $X\mathbf{d} = X\Lambda\mathbf{c} = X\Lambda L\boldsymbol{\beta} = XL^{\ddagger}L\boldsymbol{\beta} = \mathbf{0}$, since $\Lambda L = I - GX = I - X^{\ddagger}X = L^{\ddagger}L$, and $XL^{\ddagger} = 0$.

Corollary The BLU estimate of an estimable function $\boldsymbol{\theta}'\boldsymbol{\beta}$ is given by $\boldsymbol{\theta}'\hat{\boldsymbol{\beta}}^*$ where $\hat{\boldsymbol{\beta}}^* = X^{\ddagger}\mathbf{y} + L^{\ddagger}\mathbf{c}$.

Proof: By Theorem 6.3 any estimator $G\mathbf{y} + \mathbf{d}$ of $\boldsymbol{\beta}$ leads to the BLU estimate $\boldsymbol{\theta}'(G\mathbf{y} + \mathbf{d})$ if and only if $XGX = X$, $(XG)' = XG$, and $X\mathbf{d} = \mathbf{0}$.

Clearly $X^{\ddagger}$ and $L^{\ddagger}\mathbf{c}$ satisfy these conditions on G and $\mathbf{d}$ respectively.

Besides BLU estimation of estimable linear functions, an analysis must also provide an estimate of variance, and this requires in the first instance an estimate of σ^2.

It is readily shown that the sum of squares due to $\hat{\boldsymbol{\beta}}^*$ is $(\hat{\boldsymbol{\beta}}^*)'\mathbf{g} = \mathbf{y}'XK^{-1}X'\mathbf{y}$, and that the residual sum of squares is $\mathbf{y}'(I - XK^{-1}X')\mathbf{y}$. These are in correspondence with the formulae obtained in Theorem 6.6 and are therefore the same as obtained in the estimable functions approach. The estimate of σ^2 is therefore the same.

Also,

$$\begin{aligned} \mathrm{var}(\boldsymbol{\theta}'\hat{\boldsymbol{\beta}}^*) &= \sigma^2\boldsymbol{\theta}'K^{-1}SK^{-1}\boldsymbol{\theta} \\ &= \sigma^2\boldsymbol{\theta}'S^{g_2}\boldsymbol{\theta}, \end{aligned} \quad [6.22]$$

which is the same as given in Theorem 6.3, Corollary 2. Actually [6.22] reduces to $\sigma^2\boldsymbol{\theta}'K^{-1}\boldsymbol{\theta}$ by use of [6.3], as is in any case clear from the uniqueness established in §6.2.

From [3.48] we see the justification of the procedure of those practitioners who appended linear restrictions so as to make M of full rank and took the leading submatrix of $M^{-1}(\times\sigma^2)$ as $\mathrm{var}(\hat{\boldsymbol{\beta}}^*)$.

Yates and Hale calculated "the reciprocal matrix" (of S) as an extension of Fisher's C-matrix method of calculating $C = S^{-1}$, by solving $SY = I$ one vector at a time as $SY\mathbf{e}_i = \mathbf{e}_i$, where $\mathbf{e}_i$ is the ith column of I. Realizing that for S singular these equations were not consistent, they replaced the right-hand side by $B = I - L'(L^*L')^{-1}L^*$, where L^* is an orthogonal complement of S. Since $L^*K = L^*L'L$, $L'LK^{-1} = L'(L^*L')^{-1}L^*$ and $B = SK^{-1}$. (This procedure was "put on a formal basis" by Plackett (1950), except that he derived similar equations of the form $ZS = B^* = I - (L^*)'\{L(L^*)'\}^{-1}L$.) Any solution to $SY = B$ may be written as $S^{g_1}B = S^{g_1}SK^{-1}$, and, if a symmetric solution is obtained, $(S^{g_1}SK^{-1})' = K^{-1}S(S^{g_1})' = S^{g_1}SK^{-1}$, postmultiplication of which by S gives $S^{g_1}S = K^{-1}S$, i.e. $S^{g_1}SK^{-1} = K^{-1}SK^{-1}$, the required matrix. Hence Yates and Hale were probably the first to calculate a g_2-inverse $(K^{-1}SK^{-1})$, doing so with avoidance of inversion of K and remarkable economies of computation.

Scheffé (1959, p. 16) showed that the method of imposed linear restrictions is equivalent to the substitution in the model of a new set of parameters $\boldsymbol{\beta}^*$ satisfying the equations

$$\begin{bmatrix} X \\ L \end{bmatrix}\boldsymbol{\beta}^* = \begin{bmatrix} X\boldsymbol{\beta} \\ \mathbf{c} \end{bmatrix}. \quad [6.23]$$

In practice it is customary to make no change of notation, but the restrictions $L\boldsymbol{\beta}^* = \mathbf{c}$ are actually obeyed by the new parameters, not the original parameters, and the distinction indicated by the asterisk is necessary for the clarification of the theory which follows.

The matrix on the left-hand side of [6.23] has full column-rank k, and therefore, as shown in the discussion of left and right inverses in § 2.6, equations [6.23] possess the unique solution

$$\boldsymbol{\beta}^* = \begin{bmatrix} X \\ L \end{bmatrix}^g \begin{bmatrix} X\boldsymbol{\beta} \\ \mathbf{c} \end{bmatrix}$$

$$= (S + L'L)^{-1}[X' \quad L'] \begin{bmatrix} X\boldsymbol{\beta} \\ \mathbf{c} \end{bmatrix}$$

$$= (S + L'L)^{-1} S\boldsymbol{\beta} + (S + L'L)^{-1} L'\mathbf{c}$$

$$= K^{-1}S\boldsymbol{\beta} + K^{-1}L'\mathbf{c}, \qquad \textbf{[6.24]}$$

where $K = S + L'L$.

Since $K^{-1}S\boldsymbol{\beta}$ is estimable, the BLU estimate of $\boldsymbol{\beta}^*$ is given by

$$\hat{\boldsymbol{\beta}}^* = K^{-1}X'\mathbf{y} + K^{-1}L'\mathbf{c}$$

$$= X^{\ddagger}\mathbf{y} + L^{\ddagger}\mathbf{c},$$

identical with [6.16] and [6.17].

In § 6.1 estimable functions were defined in terms of homogeneous linear functions of $\boldsymbol{\beta}$, but the above slight extension offers no difficulty. In § 6.2 we also defined reparametrization of the linear model in terms of homogeneous linear transformations of $\boldsymbol{\beta}$. Accordingly $X\boldsymbol{\beta}^*$, where $\boldsymbol{\beta}^*$ is given by [6.24] but with $\mathbf{c} = \mathbf{0}$, is a reparametrization of [6.1].

The conclusion to be drawn is that, even if the parameters in the model not of full rank are envisaged as being free of constraints, by the introduction of suitable linear restrictions to solve the normal equations the practitioner is in effect reparametrizing the model and estimating parameters subject to the restrictions imposed. Finally, although $\text{var}(\hat{\boldsymbol{\beta}}^*)$ depends on L, $\text{var}(\boldsymbol{\theta}'\hat{\boldsymbol{\beta}}^*)$ does not.

In the light of the foregoing, Searle (1966, 1967) has recommended the estimability approach and the use of a g-inverse of S to express the solution of the normal equations, since in this way it is possible to avoid the practice of applying linear restrictions. In his book (1966) he also appears to recommend the use of S^{g_1} in obtaining a numerical solution.

It is difficult, however, to accept this point of view unreservedly, since, as seen above, these two aspects are so closely intertwined.

The following points may be made:

(1) Any solution to the normal equations of the form $\mathbf{b} = S^{g_1}X'\mathbf{y}$ is by Theorem 6.5 a BLU estimate of $S^{g_1}S\boldsymbol{\beta}$, and therefore of the parameters $\boldsymbol{\beta}^* = S^{g_1}S\boldsymbol{\beta}$ of a reparametrized model. These parameters and their estimates are subject to $k - p$ linearly independent linear relations $(I - S^{g_1}S)\boldsymbol{\beta}^* = \mathbf{0}$. In particular, if [6.9] is used, $\boldsymbol{\beta}^*$ is subject to $[0 \quad I]\boldsymbol{\beta}^* = \mathbf{0}$.

(2) For any solution in the form [6.8] there is, by Theorem 6.9, some set of linear restrictions which will give the same result.

(3) As regards the numerical solution of the normal equations, the computed S^{g_1} will probably be [6.9] or, in statistical problems, $(S + L'L)^{-1}$. An enlightened computor will wish to know how this relates to the linear restrictions approach.

(4) If it is desired to obtain the BLU estimate of $\boldsymbol{\beta}^* = T\boldsymbol{\beta}$, a set of estimable linear functions determinable as in [6.24] by an obvious set of natural restrictions $L\boldsymbol{\beta} = \mathbf{0}$, it is preferable to compute $\hat{\boldsymbol{\beta}}^*$ directly from [6.16] as $K^{-1}X'\mathbf{y}$ than as $TS^{g_1}X'\mathbf{y}$, where S^{g_1} is some "anonymous" g_1-inverse of S. In this reparametrization X^* can be taken as X, and the computation of $K^{-1}X'$ avoids the difficulties in respect of recognition of zero pivotal elements or rows and columns mentioned in §4.1.

6.5 Linear models with *a priori* linear constraints

In this section the problem of estimating linear functions of the parameters in the model $\mathbf{y} = X\boldsymbol{\beta} + \boldsymbol{\varepsilon}$, where X is $n \times k$ of rank $p < k$, $\text{var}(\mathbf{y}) = \sigma^2 I$, and $\boldsymbol{\beta}$ is subject to the linear constraints $L\boldsymbol{\beta} = \mathbf{c}$, is considered. The essential difference between this case and that of §6.3 is that the restrictions are assumed to be known in advance, and are not merely imposed for the purpose of reparametrization or of obtaining a unique solution to the normal equations. It is assumed that the constraints are consistent and that L is a $q \times k$ matrix of rank $s \leqslant q \leqslant k$. *However, no assumptions are made about L in relation to the row-space of X.*

The minimization of the constrained sum of squared residuals leads to equations of the same form as [6.13], namely

$$\begin{bmatrix} S & L' \\ L & 0 \end{bmatrix} \begin{bmatrix} \hat{\boldsymbol{\beta}} \\ \boldsymbol{\lambda} \end{bmatrix} = \begin{bmatrix} \mathbf{g} \\ \mathbf{c} \end{bmatrix},$$

where as before $S = X'X$, $\mathbf{g} = X'\mathbf{y}$, and $\boldsymbol{\lambda}$ is a vector of Lagrange multipliers. For convenience we write these equations as

$$M\boldsymbol{\alpha} = \mathbf{h}. \qquad [6.25]$$

In the following two theorems the condition for a linear function of the parameters to be estimable will be derived, and the BLU estimate of an estimable linear function is obtained. We first derive certain relationships used in proving the theorems. The matrix M is such that it has as a particular g_1-inverse the matrix [3.43], which we denote by $G = \begin{bmatrix} G_{11} & G_{12} \\ G_{21} & G_{22} \end{bmatrix}$ with the partitioning corresponding to that of [3.43].

After simplification utilizing [3.38] and [3.42], the product GM reduces to

$$GM = \begin{bmatrix} K^{g_1}K & 0 \\ 0 & R^{g_1}R \end{bmatrix}, \qquad [6.26]$$

whence

$$G_{11}S + G_{12}L = K^{g_1}K \qquad [6.27]$$

and

$$G_{11}L' = \mathbf{0}. \qquad [6.28]$$

THEOREM 6.10 A linear function, $\boldsymbol{\theta}'\boldsymbol{\beta}$, of the constrained parameters is estimable if and only if $\boldsymbol{\theta}'K^{g_1}K = \boldsymbol{\theta}'$, where $K = S + L'L$.

Proof: If $\boldsymbol{\theta}'\boldsymbol{\beta}$ is estimable, there exists $\mathbf{a}_1'\mathbf{y} + d$, some linear estimator of $\boldsymbol{\theta}'\boldsymbol{\beta}$, such that

$$\begin{aligned} \mathrm{E}(\mathbf{a}_1'\mathbf{y} + d) &= \mathbf{a}_1'X\boldsymbol{\beta} + d \\ &\equiv \boldsymbol{\theta}'\boldsymbol{\beta} + \mathbf{a}_2'(\mathbf{c} - L\boldsymbol{\beta}), \end{aligned}$$

where $\mathbf{a}_2'$ is a Lagrangian multiplier. This implies that

$$\mathbf{a}_1'X + \mathbf{a}_2'L = \boldsymbol{\theta}' \qquad [6.29]$$

and

$$d = \mathbf{a}_2'\mathbf{c}. \qquad [6.30]$$

Therefore $\boldsymbol{\theta}'K^{g_1}K = (\mathbf{a}_1'X + \mathbf{a}_2'L)K^{g_1}K = \mathbf{a}_1'X + \mathbf{a}_2'L$ (by [3.37] and [3.39]) $= \boldsymbol{\theta}'$.

Conversely, if $\boldsymbol{\theta}'K^{g_1}K = \boldsymbol{\theta}'$ and $\hat{\boldsymbol{\beta}}$ is as given by any solution to [6.25],

$$\begin{aligned} \boldsymbol{\theta}'\hat{\boldsymbol{\beta}} &= \boldsymbol{\theta}'\{G_{11}X'\mathbf{y} + G_{12}\mathbf{c} + (I - K^{g_1}K)\mathbf{z}\} \quad \text{(where } \mathbf{z} \text{ is arbitrary)}, \\ &= \boldsymbol{\theta}'(G_{11}X'\mathbf{y} + G_{12}\mathbf{c}), \qquad [6.31] \end{aligned}$$

so that

$$\begin{aligned} \mathrm{E}(\boldsymbol{\theta}'\hat{\boldsymbol{\beta}}) &= \boldsymbol{\theta}'(G_{11}S + G_{12}L)\boldsymbol{\beta} \\ &= \boldsymbol{\theta}'K^{g_1}K\boldsymbol{\beta} \quad \text{by [6.27]} \\ &= \boldsymbol{\theta}'\boldsymbol{\beta}, \end{aligned}$$

i.e. $\boldsymbol{\theta}'\boldsymbol{\beta}$ is estimable.

THEOREM 6.11 If $\hat{\boldsymbol{\beta}}$ is as given by any solution to [6.25] then:

(a) $\boldsymbol{\theta}'\hat{\boldsymbol{\beta}}$ is unique;

(b) $\text{var}(\boldsymbol{\theta}'\hat{\boldsymbol{\beta}}) = \sigma^2\boldsymbol{\theta}'G_{11}\boldsymbol{\theta}$, where G_{11} is the leading submatrix of G, the g_1-inverse of M in [3.43];

(c) $\boldsymbol{\theta}'\hat{\boldsymbol{\beta}}$ is the BLU estimate of the estimable linear function $\boldsymbol{\theta}'\boldsymbol{\beta}$.

Proof:

(a) Let $\boldsymbol{\Phi}'\boldsymbol{\alpha}$ be a linear combination of the elements of $\boldsymbol{\alpha}$, any solution to [6.25]. By Theorem 1.6, Corollary 6, $\boldsymbol{\Phi}'\boldsymbol{\alpha}$ is unique if and only if $\boldsymbol{\Phi}' = \boldsymbol{\Phi}'GM$. Hence, if $\boldsymbol{\Phi}' = [\boldsymbol{\theta}' \quad \mathbf{0}']$, so that $\boldsymbol{\Phi}'\boldsymbol{\alpha} = \boldsymbol{\theta}'\hat{\boldsymbol{\beta}}$, it follows from [6.26] that $\boldsymbol{\theta}'\hat{\boldsymbol{\beta}}$ is unique if and only if $\boldsymbol{\theta}'K^{g_1}K = \boldsymbol{\theta}'$, which must hold since $\boldsymbol{\theta}'\boldsymbol{\beta}$ is estimable (Theorem 6.10).

(b) $\text{var}(\boldsymbol{\theta}'\hat{\boldsymbol{\beta}}) = \text{var}\{\boldsymbol{\theta}'(G_{11}X'\mathbf{y})\}$ by [6.31],

$$= \sigma^2\boldsymbol{\theta}'G_{11}SG_{11}'\boldsymbol{\theta}.$$

Now by [6.27],

$$\boldsymbol{\theta}' = \boldsymbol{\theta}'K^{g_1}K = \boldsymbol{\theta}'(G_{11}S + G_{12}L),$$

so that

$$\boldsymbol{\theta}'G_{11}' = \boldsymbol{\theta}'G_{11}SG_{11}' \quad \text{by [6.28]},$$

and $\text{var}(\boldsymbol{\theta}'\hat{\boldsymbol{\beta}}) = \sigma^2\boldsymbol{\theta}'G_{11}\boldsymbol{\theta}$. That $\text{var}(\boldsymbol{\theta}'\hat{\boldsymbol{\beta}})$ is unique follows from statistical considerations. An algebraic proof based on the properties of M^{g_1} can be constructed but is complex and will not be presented.

(c) As already seen in Theorem 6.10, $\boldsymbol{\theta}'\hat{\boldsymbol{\beta}} = \boldsymbol{\theta}'(G_{11}\mathbf{g} + G_{12}\mathbf{c})$ is an unbiased estimate of $\boldsymbol{\theta}'\boldsymbol{\beta}$, and from (b) its variance is $\sigma^2\boldsymbol{\theta}'G_{11}\boldsymbol{\theta}$. Now, in accordance with [6.30], let $\mathbf{a}_1'\mathbf{y} + \mathbf{a}_2'\mathbf{c}$ be any other unbiased estimate with variance $\sigma^2\mathbf{a}_1'\mathbf{a}_1$. Since this estimate is unbiased, [6.29] holds, i.e.

$$\mathbf{a}_1'X = \boldsymbol{\theta}' - \mathbf{a}_2'L$$

and

$$\mathbf{a}_1'XG_{11}' = \boldsymbol{\theta}'G_{11}',$$

so that $(\mathbf{a}_1' - \boldsymbol{\theta}'G_{11}X')(\mathbf{a}_1' - \boldsymbol{\theta}'G_{11}X')' = \mathbf{a}_1'\mathbf{a}_1 - \boldsymbol{\theta}'G_{11}SG_{11}'\boldsymbol{\theta}$. It now follows that the variance of the alternative estimate can never be less than $\text{var}(\boldsymbol{\theta}'\hat{\boldsymbol{\beta}})$, i.e. $\boldsymbol{\theta}'\hat{\boldsymbol{\beta}}$ is the BLU estimate.

The expressions for $\boldsymbol{\theta}'\hat{\boldsymbol{\beta}}$ and $\text{var}(\boldsymbol{\theta}'\hat{\boldsymbol{\beta}})$ given above reduce to those presented by Dwyer (1958) for the special case when S is non-singular and $\text{r}(L) = q$.

Rao (1965, pp. 189–91) gives proofs of (b) and (c), but without explicit expressions for $\hat{\boldsymbol{\beta}}$ and $\operatorname{var}(\boldsymbol{\theta}'\hat{\boldsymbol{\beta}})$. In these circumstances the validity of his proof of (b) is not clear to the authors.

Chipman (1964) and Goldman & Zelen (1964) have also considered the problem of linear estimation subject to prior restrictions, but their methods involve the separation of the rows of L into those belonging to the row-space of S and those belonging to the complementary space.

6.6 Some aspects of partitioned linear models

In this section the situation where special groups of parameters in the model [6.1] are of particular interest is considered. Let X be partitioned as $X = [X_1 \quad X_2]$, where X_i $(i = 1,2)$ is $n \times k_i$ $(k_1 + k_2 = k)$, with a conformable partitioning of $\boldsymbol{\beta}$ as $\begin{bmatrix} \boldsymbol{\beta}_1 \\ \boldsymbol{\beta}_2 \end{bmatrix}$. The model may thus be written as

$$\mathbf{y} = X_1\boldsymbol{\beta}_1 + X_2\boldsymbol{\beta}_2 + \boldsymbol{\varepsilon}. \qquad [6.32]$$

The first problem which will be considered is that of obtaining a solution to the normal equations by handling subsets of these equations. This is achieved by obtaining a set of equations for $\mathbf{b}_2$ eliminating $\mathbf{b}_1$.

The normal equations $S\mathbf{b} = \mathbf{g}$ may be written in partitioned form as

$$\begin{bmatrix} S_{11} & S_{12} \\ S_{21} & S_{22} \end{bmatrix} \begin{bmatrix} \mathbf{b}_1 \\ \mathbf{b}_2 \end{bmatrix} = \begin{bmatrix} \mathbf{g}_1 \\ \mathbf{g}_2 \end{bmatrix}. \qquad [6.33]$$

Now premultiplication of [6.33] by $\begin{bmatrix} I & 0 \\ -C & I \end{bmatrix}$ gives

$$\begin{bmatrix} S_{11} & S_{12} \\ S_{21} - CS_{11} & S_{22} - CS_{12} \end{bmatrix} \begin{bmatrix} \mathbf{b}_1 \\ \mathbf{b}_2 \end{bmatrix} = \begin{bmatrix} \mathbf{g}_1 \\ \mathbf{g}_2 - C\mathbf{g}_1 \end{bmatrix},$$

so that a set of equations for $\mathbf{b}_2$ eliminating $\mathbf{b}_1$ will be obtained if C satisfies the equations

$$CS_{11} = S_{21}. \qquad [6.34]$$

Since the column-space of S_{12} is contained in that of S_{11}, a solution for C clearly exists.

The equations for $\mathbf{b}_2$ eliminating $\mathbf{b}_1$ are therefore

$$(S_{22} - CS_{12})\mathbf{b}_2 = \mathbf{g}_2 - C\mathbf{g}_1. \qquad [6.35]$$

Now by Theorem 1.6, Corollary 3, $C = S_{21}S_{11}^{g_1}$ is a solution to [6.34]. However, since S is singular there are many possible solutions, and it is necessary to establish that [6.35] is invariant under choice of solution of [6.34], i.e. it must be shown that CS_{12} and $C\mathbf{g}_1$ are unique. The uniqueness of CS_{12} follows immediately from Theorem 1.6, Corollary 5, since S_{12} is contained in the column-space of S_{11}. A similar argument shows that $C\mathbf{g}_1 = CX_1'\mathbf{y}$ is unique. Thus the substitution $C = S_{21}S_{11}^{g_1}$ in [6.35] gives

$$(S_{22} - S_{21}S_{11}^{g_1}S_{12})\mathbf{b}_2 = \mathbf{g}_2 - S_{21}S_{11}^{g_1}\mathbf{g}_1$$

or

$$S_{22}^*\mathbf{b}_2 = \mathbf{g}_2^*. \qquad [6.36]$$

The process whereby equations [6.36] were derived is a generalization of the "sweep-out" process of obtaining reduced normal equations described by Anderson & Bancroft (1952, p. 280).

A solution for $\mathbf{b}$ from the full $k \times k$ set of normal equations can therefore be obtained in two stages – by first solving the $k_2 \times k_2$ set of equations [6.36] and then using $\mathbf{b}_2 = (S_{22}^*)^{g_1}\mathbf{g}_2^*$ to obtain a solution to

$$S_{11}\mathbf{b}_1 = \mathbf{g}_1 - S_{12}\mathbf{b}_2. \qquad [6.37]$$

If $\mathbf{b}_1^*$ represents a solution to the equations $S_{11}\mathbf{b}_1 = \mathbf{g}_1$, i.e. $\mathbf{b}_1^* = S_{11}^{g_1}\mathbf{g}_1$, then a solution to [6.37] can be written in the form

$$\mathbf{b}_1 = \mathbf{b}_1^* - S_{11}^{g_1}S_{12}\mathbf{b}_2. \qquad [6.38]$$

It can now be seen that this two-stage technique is admirably suited to cope with the general problems of addition and deletion of parameters. Consider, for example, the problem of augmenting the model $\mathbf{y} = X_1\boldsymbol{\beta}_1 + \boldsymbol{\varepsilon}$ with k_2 parameters $\boldsymbol{\beta}_2$. It is clear that there is no need to solve the full set of k normal equations *de novo*. The matrix $S_{11}^{g_1}$ which would have been computed in solving for $\mathbf{b}_1^*$ may be used to construct equations [6.36], which in turn yield a solution for $\mathbf{b}_2$, and thus the adjustment required in [6.38].

Estimability and linear restrictions in partitioned models

Thus far the discussion has been confined to a stepwise solution of the normal equations. Some attention will now be paid to the conditions for estimability and to the effect of linear restrictions in partitioned models.

Zyskind (1964) has shown that the reduced normal equations [6.36] contain all the required information regarding best linear unbiased estimation of estimable linear functions of $\boldsymbol{\beta}_2$. This may be seen as follows.

THEOREM 6.12

(a) $\boldsymbol{\theta}'\boldsymbol{\beta}_2$ is estimable if and only if

$$\boldsymbol{\theta}'(S_{22}^*)^{g_1}S_{22}^* = \boldsymbol{\theta}'. \qquad [6.39]$$

(b) The BLU estimate of an estimable linear function $\boldsymbol{\theta}'\boldsymbol{\beta}_2$ is $\boldsymbol{\theta}'\mathbf{b}_2$, where $\mathbf{b}_2$ is a solution to [6.36].

(c) $\text{var}(\boldsymbol{\theta}'\mathbf{b}_2) = \sigma^2\boldsymbol{\theta}'(S_{22}^*)^{g_1}\boldsymbol{\theta}$.

Proof: By [6.4], a linear function $\boldsymbol{\theta}'\boldsymbol{\beta}_2 = [\mathbf{0}' \quad \boldsymbol{\theta}']\begin{bmatrix}\boldsymbol{\beta}_1\\ \boldsymbol{\beta}_2\end{bmatrix}$ is estimable

if and only if $[\mathbf{0}' \quad \boldsymbol{\theta}']S^{g_1}S = [\mathbf{0}' \quad \boldsymbol{\theta}']$. Now if S is partitioned as in equation [6.33], it has as a particular g_1-inverse the matrix of [3.33] in which $Z = S_{22}^*$, whence

$$S^{g_1}S = \begin{bmatrix} S_{11}^{g_1}S_{11} & S_{11}^{g_1}S_{12}[I - (S_{22}^*)^{g_1}S_{22}^*] \\ 0 & (S_{22}^*)^{g_1}S_{22}^* \end{bmatrix}.$$

It follows therefore that $\boldsymbol{\theta}'\boldsymbol{\beta}_2$ is estimable if and only if [6.39] holds.

A proof of (b) follows from Theorem 6.3, Corollary 3, since the BLU estimate of $\boldsymbol{\theta}'\boldsymbol{\beta}_2$ is $\boldsymbol{\theta}'\mathbf{b}_2$, where $\mathbf{b}_2$ is as given by any solution to the full set of normal equations, i.e. where $\mathbf{b}_2$ is any solution to the reduced normal equations.

Theorem 6.3, Corollary 2, and the expression [3.33] for S^{g_1} combine to prove (c).

It is interesting to note that $\mathbf{b}_2$ is unaffected by any linear constraints upon $\boldsymbol{\beta}_1$ and $\mathbf{b}_1$. Suppose that q_1 linear restrictions $L_1\boldsymbol{\beta}_1 = \mathbf{c}_1$, where L_1 is complementary to X_1, are imposed upon $\boldsymbol{\beta}_1$. For this case the normal equations may be written as (cf. [6.14])

$$\left[\begin{array}{cc|c} S_{11} & L_1' & S_{12} \\ L_1 & 0 & 0 \\ \hline S_{21} & 0 & S_{22} \end{array}\right]\left[\begin{array}{c} \mathbf{b}_1 \\ \boldsymbol{\lambda}_1 \\ \hline \mathbf{b}_2 \end{array}\right] = \left[\begin{array}{c} \mathbf{g}_1 \\ \mathbf{c}_1 \\ \hline \mathbf{g}_2 \end{array}\right],$$

or as

$$\begin{bmatrix} M_{11} & M_{12} \\ M_{21} & M_{22} \end{bmatrix}\begin{bmatrix} \boldsymbol{\alpha}_1 \\ \mathbf{b}_2 \end{bmatrix} = \begin{bmatrix} \mathbf{h}_1 \\ \mathbf{g}_2 \end{bmatrix}. \qquad [6.40]$$

Now, since L_1 is complementary to X_1

$$M_{11}^{g} = \begin{bmatrix} K_1^{-1}S_{11}K_1^{-1} & K_1^{-1}L_1' \\ L_1K_1^{-1} & 0 \end{bmatrix},$$

where $K_1^{-1} = (S_{11} + L_1'L_1)^{-1}$. Hence by Theorem 2.19 $M_{21}M_{11}^{g} = [S_{21}K_{11}^{-1} \quad 0]$.

Premultiplication of [6.40] by

$$\begin{bmatrix} I & 0 \\ -M_{21}M_{11}^{g} & I \end{bmatrix}$$

gives
$$\begin{bmatrix} M_{11} & M_{12} \\ 0 & M_{22} - M_{21}M_{11}^{g}M_{12} \end{bmatrix}\begin{bmatrix} \boldsymbol{\alpha}_1 \\ \mathbf{b}_2 \end{bmatrix} = \begin{bmatrix} \mathbf{h}_1 \\ \mathbf{g}_2 - M_{21}M_{11}^{g}\mathbf{h}_1 \end{bmatrix}.$$

Now $M_{22} - M_{21}M_{11}^{g}M_{12} = S_{22} - S_{21}K_{11}^{-1}S_{12} = S_{22}^{*}$ by Theorem 2.12, and similarly $\mathbf{g}_2 - M_{21}M_{11}^{g}\mathbf{h}_1 = \mathbf{g}_2 - S_{21}K_{11}^{-1}\mathbf{g}_1 = \mathbf{g}_2^{*}$. Hence the reduced normal equations for $\mathbf{b}_2$ are identical with [6.36].

Tests of hypotheses with partitioned models

Of main interest here is the test of the null hypothesis $\boldsymbol{\beta}_2 = \mathbf{0}$. This test is, of course, used to decide whether to delete from the model the terms involving $\boldsymbol{\beta}_2$, or to retain them.

By Theorem 6.6 (a) the sum of squares due to fitting the full model [6.1] is $\mathbf{b}'S\mathbf{b}$, and that due to fitting the truncated model (with $\boldsymbol{\beta}_2 = \mathbf{0}$) is $(\mathbf{b}_1^{*})'S_{11}\mathbf{b}_1^{*}$. Hence the sum of squares due to the hypothesis (due to fitting $\boldsymbol{\beta}_2$ after $\boldsymbol{\beta}_1$) is $\mathbf{b}'S\mathbf{b} - (\mathbf{b}_1^{*})'S_{11}\mathbf{b}_1^{*}$. This may be written in partitioned form as

$$[\mathbf{b}_1' \quad \mathbf{b}_2'] \begin{bmatrix} S_{11} & S_{12} \\ S_{21} & S_{22} \end{bmatrix}\begin{bmatrix} \mathbf{b}_1 \\ \mathbf{b}_2 \end{bmatrix} - (\mathbf{b}_1^{*})'S_{11}\mathbf{b}_1^{*}$$

$$= \mathbf{b}_1'S_{11}\mathbf{b}_1 + \mathbf{b}_1'S_{12}\mathbf{b}_2 + \mathbf{b}_2'S_{21}\mathbf{b}_1 + \mathbf{b}_2'S_{22}\mathbf{b}_2 - (\mathbf{b}_1^{*})'S_{11}\mathbf{b}_1^{*}. \qquad \textbf{[6.41]}$$

But from [6.38] $\mathbf{b}_1^{*} = \mathbf{b}_1 + S_{11}^{g}S_{12}\mathbf{b}_2$. Hence by [2.5] and [2.6]

$$(\mathbf{b}_1^{*})'S_{11}\mathbf{b}_1^{*} = \mathbf{b}_1'S_{11}\mathbf{b}_1 + \mathbf{b}_1'S_{12}\mathbf{b}_2 + \mathbf{b}_2'S_{21}\mathbf{b}_1 + \mathbf{b}_2'S_{21}S_{11}^{g}S_{12}\mathbf{b}_2. \qquad \textbf{[6.42]}$$

Substitution of [6.42] in [6.41] gives

$$\mathbf{b}_2'S_{22}\mathbf{b}_2 - \mathbf{b}_2'S_{21}S_{11}^{g}S_{12}\mathbf{b}_2$$
$$= \mathbf{b}_2'S_{22}^{*}\mathbf{b}_2.$$

If now we substitute for $\mathbf{b}_2$ the general solution to [6.36] obtainable from Theorem 1.6, Corollary 4, the arbitrary part vanishes, leaving $\mathbf{b}_2' S_{22}^* \mathbf{b}_2 = (\mathbf{g}_2^*)'(S_{22}^*)^{g_1} S_{22}^* (S_{22}^*)^{g_1} \mathbf{g}_2^*$. Now write $(I - X_1 S_{11}^{g_1} X_1')X_2 = Y$. Then $S_{22}^* = Y'Y$ and $\mathbf{g}_2^* = Y'\mathbf{y}$, and

$$\begin{aligned} \mathbf{b}_2' S_{22}^* \mathbf{b}_2 &= \mathbf{y}'Y(Y'Y)^{g_1} Y'Y(Y'Y)^{g_1} Y'\mathbf{y} \\ &= \mathbf{y}'Y(Y'Y)^{g_1} Y'\mathbf{y} \quad \text{by [2.5]} \qquad [6.43] \\ &= (\mathbf{g}_2^*)' S_{22}^* \mathbf{g}_2^*, \end{aligned}$$

which is unique.

From the foregoing and Theorem 6.6 it is seen that the total sum of squares has been partitioned according to

$$\mathbf{y}'\mathbf{y} = q_1 + q_2 + q_3,$$

where q_1 and q_2 are the quadratic forms [6.42] and [6.43] adding to $\mathbf{b}'S\mathbf{b}$. and q_3 is the $SS(E)$ of Theorem 6.6. As quadratic forms in $\mathbf{y}$, q_1, q_2, and q_3 have matrices $X_1 S_{11}^{g_1} X_1'$, $Y(Y'Y)^{g_1}Y'$, and $I - XS^{g_1}X'$, respectively, with ranks equal to $\mathrm{r}(X_1) = p_1$ (say), $\mathrm{r}(Y) = \mathrm{r}(S_{22}^*)$, and $n - p$. From the reduction

$$\begin{bmatrix} I & 0 \\ -S_{21}S_{11}^{g_1} & I \end{bmatrix} S \begin{bmatrix} I & -S_{11}^{g_1}S_{12} \\ 0 & I \end{bmatrix} = \begin{bmatrix} S_{11} & 0 \\ 0 & S_{22}^* \end{bmatrix}$$

(cf. [3.24]) it is seen that $\mathrm{r}(S_{22}^*) = p - p_1$. Hence the ranks of q_1, q_2, and q_3 sum to n, the rank of $\mathbf{y}'\mathbf{y}$, and by an extension of Cochran's theorem it follows that if $\mathbf{y}$ is $\mathrm{N}(\boldsymbol{\mu}, \sigma^2 I)$, q_1, q_2, and q_3 have independent noncentral (possibly central) χ^2 distributions with respective degrees of freedom p_1, $p - p_1$, and $n - p$.

Under the null hypothesis $\boldsymbol{\beta}_2 = \mathbf{0}$, $\boldsymbol{\mu}$ is equal to $X_1\boldsymbol{\beta}_1$, and the noncentrality parameters are respectively (by Theorem 5.3):

For q_1 : $\boldsymbol{\beta}_1' X_1' X_1 S_{11}^{g_1} X_1' X_1 \boldsymbol{\beta}_1/\sigma^2 = \boldsymbol{\beta}_1' S_{11} \boldsymbol{\beta}_1/\sigma^2$.

For q_2 : 0, since $X_1'Y = 0$ by [2.6].

For q_3 : 0, since $(XS^{g_1}X')X = X$, i.e. $XS^{g_1}X'X_1 = X_1$.

Under the alternative hypothesis $\boldsymbol{\beta}_2 \neq \mathbf{0}$, q_2 has noncentrality parameter $\boldsymbol{\beta}'X'Y(Y'Y)^{g_1}Y'X\boldsymbol{\beta}/\sigma^2 = \boldsymbol{\beta}_2' S_{22}^* (S_{22}^*)^{g_1} S_{22}^* \boldsymbol{\beta}_2/\sigma^2 = \boldsymbol{\beta}_2' S_{22}^* \boldsymbol{\beta}_2/\sigma^2$, since

$$X'Y = \begin{bmatrix} 0 \\ S_{22}^* \end{bmatrix}.$$

Hence the appropriate test of significance is performed by means of

$$F = \frac{q_2}{p - p_1} \bigg/ \frac{q_3}{n - p},$$

which on the null hypothesis follows the F-distribution with $p - p_1$ and $n - p$ degrees of freedom.

Various other tests may be described using the results developed in this section. For example, using the methods of § 6.2 a test of the hypothesis $H_0 : U\boldsymbol{\beta}_1 = \mathbf{z}$, where $U\boldsymbol{\beta}_1$ is a set of estimable linear functions, may be derived.

Analysis of covariance

In this context the rows of X_2 contain the corresponding values of specified concomitant variates, and model [6.32] is called the "analysis of covariance model", while the model $\mathbf{y} = X_1\boldsymbol{\beta}_1 + \boldsymbol{\varepsilon}$ is the corresponding analysis of variance model. It is reasonable to assume in this case that $\mathrm{r}[X_1 \quad X_2] = \mathrm{r}(X_1) + \mathrm{r}(X_2)$, and that $\mathrm{r}(X_2) = k_2$, in which case $\mathrm{r}(S_{22}^*) = \mathrm{r}(X) - \mathrm{r}(X_1) = \mathrm{r}(X_2)$, i.e. S_{22}^* is nonsingular (cf. § 3.3, below [3.30]).

If $\boldsymbol{\theta}'\boldsymbol{\beta}_1$ is estimable with respect to the analysis of variance model, i.e. if $\boldsymbol{\theta}' S_{11}^{g_1} S_{11} = \boldsymbol{\theta}'$, then it is easily seen from the expression for $S^{g_1}S$ given in Theorem 6.12 with $(S_{22}^*)^{g_1} = (S_{22}^*)^{-1}$ that $[\boldsymbol{\theta}' \quad \mathbf{0}'] S^{g_1}S = [\boldsymbol{\theta}' \quad \mathbf{0}']$, i.e. $\boldsymbol{\theta}'\boldsymbol{\beta}_1$ is also estimable under the analysis of covariance model. The BLU estimate of $\boldsymbol{\theta}'\boldsymbol{\beta}_1$ when the covariates are included is

$$\boldsymbol{\theta}'\mathbf{b}_1 = \boldsymbol{\theta}'\mathbf{b}_1^* - \boldsymbol{\theta}' S_{11}^{g_1} S_{12}\mathbf{b}_2$$

by [6.38].

It follows also from Theorem 6.12, with S_{22}^* nonsingular, that $\mathbf{b}_2$ is the BLU estimate of the vector of regression coefficients of the covariates.

6.7 Some statistical interpretations of g-inverses in linear estimation procedures

In § 6.2 it was shown that for $G\mathbf{y}$ to be a least-squares solution to the observational equations $X\mathbf{b} = \mathbf{y}$, G must be a g-inverse of X satisfying conditions (1) and (3). We present here two similar theorems involving the g-inverse of X.

The first theorem is due to Bjerhammar (1958), although the proof is taken from Albert & Sittler (1966).

THEOREM 6.13 A necessary and sufficient condition that $G\mathbf{y}$ should be a minimum norm least-squares solution to $X\mathbf{b} = \mathbf{y}$ is $G = X^g$.

Proof: Let $\mathbf{b} = X^g\mathbf{y}$ and let $\mathbf{b}_0$ be any other least-squares solution. Further, let $\mathbf{b}_0$ be expressed as $\mathbf{b}_0 = S^gS\mathbf{b}_0 + (I - S^gS)\mathbf{b}_0$. Then, since $(S^gS\mathbf{b}_0)'(I - S^gS)\mathbf{b}_0 = 0$,

$$\mathbf{b}_0'\mathbf{b}_0 = (S^gS\mathbf{b}_0)'(S^gS\mathbf{b}_0) + [(I - S^gS)\mathbf{b}_0]'[(I - S^gS\mathbf{b}_0)] \geqslant (S^gS\mathbf{b}_0)'(S^gS\mathbf{b}_0).$$

Now, since $\mathbf{b}$ and $\mathbf{b}_0$ are both solutions to the normal equations, $S\mathbf{b} = S\mathbf{b}_0 = X'\mathbf{y}$, or $S^gS\mathbf{b} = S^gS\mathbf{b}_0$. Since $\mathbf{b} = X^g\mathbf{y} = S^gX'\mathbf{y}$,

$$\begin{aligned} S^gS\mathbf{b}_0 &= S^gSS^gX'\mathbf{y} \\ &= S^gX'\mathbf{y} \\ &= \mathbf{b}, \end{aligned}$$

whence $\mathbf{b}'\mathbf{b} \leqslant \mathbf{b}_0'\mathbf{b}_0$.

The second theorem is a special case of a result proved by Chipman (1964). However, Chipman mentions that his result is, in effect, the same as that derived by Penrose (1956), although Penrose used an alternative approach.

In the linear model of full rank (X of rank k in [6.1]), the condition $GX = I$ is necessary and sufficient for $G\mathbf{y}$ to be an unbiased estimate of $\boldsymbol{\beta}$. When $\mathrm{r}(X) = p < k$, this condition is impossible of fulfilment, and the present theorem concerns best linear minimum-bias estimators, i.e. estimators which minimize the "bias matrix" $(I - GX)(I - GX)'$ and have minimum variance.

THEOREM 6.14 A necessary and sufficient condition for $G\mathbf{y}$ to be the best linear minimum-bias estimate of $\boldsymbol{\beta}$ is that G should be the g-inverse of X.

Proof: The proof is divided into two parts. The first part is concerned only with minimum bias.

(a) A necessary and sufficient condition that $(I - GX)(I - GX)'$ should be a minimum is

$$GX = X^gX, \qquad [6.44]$$

or, equivalently, that

$$XGX = X \text{ and } (GX)' = GX. \qquad [6.45]$$

This may be seen as follows:

Let $(I - GX)$ be expressed as

$$(I - GX) = (X^gX - GX) + (I - X^gX).$$

Then, since $X^gX(I - X^gX) = 0$,

$$(I - GX)(I - GX)' = (X^g X - GX)(X^g X - GX)' + (I - X^g X).$$

Familiar arguments now lead to the condition [6.44]. The minimum bias is therefore $I - X^g X$. Furthermore, it is readily verified that [6.44] and [6.45] are equivalent.

(b) As regards the property of minimum variance, if $GX = X^g X$ then $GXX^g = X^g$. Let G be expressed as $G = X^g + (G - X^g)$; then

$$\text{var}(G\mathbf{y}) = \sigma^2 GG' = \sigma^2 [X^g(X^g)' + (G - X^g)(G - X^g)'],$$

since $GXX^g = X^g$. The second term vanishes if $G = X^g$.

Chipman then proceeds to show that $G\mathbf{y}$ minimizes $(I - GX)(I - GX)'$ if and only if it is conditionally unbiased, subject to $L\boldsymbol{\beta} = \mathbf{0}$, where L is any orthogonal complement of X.

The condition for minimum bias, [6.45], is that G is a g-inverse of X satisfying conditions (1) and (4) (cf. Theorem 6.3). Actually [6.44] is more basically $GX = X^{g_3^*} X$ (cf. [6.7]), but $X^{g_3^*} X = X^g X$ (Theorem 2.13, Corollary 3). Also [6.44] and [6.45] are both equivalent to $GXX' = X'$ (Penrose, 1955), a result to be compared with Theorem 6.3, Corollary 1 (c).

Chapter 7

THE LINEAR MODEL WITH SINGULAR VARIANCE MATRIX

7.1 Introduction

In this chapter we consider the extension of the results on linear estimation to the case where the vector of observations has variance matrix V (assumed known) at least positive semidefinite. The model is, therefore, $\mathbf{y} = X\boldsymbol{\beta} + \boldsymbol{\varepsilon}$, where X is $n \times k$ of rank $p < k$, and $\boldsymbol{\varepsilon}$ has zero mean and variance matrix V of rank $r \leqslant n$.

First of all it must be noted that the various equivalent conditions [6.2], [6.3], and [6.4] for $\boldsymbol{\theta}'\boldsymbol{\beta}$ to be estimable under model [6.1], having relation only to the mean of $\mathbf{y}$, are unaffected by the change of variance matrix.

Aitken (1934, 1945) showed that for $p = k$ and $r = n$ the BLU estimate of $\boldsymbol{\beta}$ is given by the solution to the generalized normal equations $X'V^{-1}X\mathbf{b} = X'V^{-1}\mathbf{y}$, namely $\mathbf{b} = (X'V^{-1}X)^{-1}X'V^{-1}\mathbf{y}$. In the case considered by Aitken all linear functions of $\boldsymbol{\beta}$ are estimable, but it is not difficult to extend the results of Chapter 6 to show that for $p < k$ and V positive definite the BLU estimate of $\boldsymbol{\theta}'\boldsymbol{\beta}$ is given by $\boldsymbol{\theta}'\mathbf{b}$, where $\mathbf{b}$ is any solution to $X'V^{-1}X\mathbf{b} = X'V^{-1}\mathbf{y}$. The question of the existence of similar equations when V is singular is relevant since, as Zyskind & Martin (1967) point out, it is not only a matter of a general unified formulation of the theory of the linear model, but also because problems involving singular variance matrices do arise naturally in statistical theory.

It would appear that the case where V may be singular was first discussed by Goldman & Zelen (1964), who approached the problem by transforming the model to one with variance matrix $\sigma^2 I$ and specified restrictions upon $\boldsymbol{\beta}$. Rohde (1964, pp. 87–9), in discussing the case V singular, followed Goldman & Zelen's methods. The recent papers of Khatri (1968) and Mitra & Rao (1968a) also discuss the case V singular from the viewpoint of least-squares theory with constraints on the parameters.

Zyskind & Martin (1967), however, have approached the problem differently. These authors have investigated the construction of a class of g_1-inverses of V, namely V^*, such that a BLU estimate of an estimable linear function $\boldsymbol{\theta}'\boldsymbol{\beta}$ is given by $\boldsymbol{\theta}'\mathbf{b}$, where $\mathbf{b}$ is a solution to $X'V^*X\mathbf{b} = X'V^*\mathbf{y}$.

7.2 The Goldman–Zelen method

We begin by stating the principal result of this section.

THEOREM 7.1 The BLU estimate of an estimable linear function $\boldsymbol{\theta}'\boldsymbol{\beta}$ in the linear model $E(\mathbf{y}) = X\boldsymbol{\beta}$, $\text{var}(\mathbf{y}) = V$, is $\boldsymbol{\theta}'\hat{\boldsymbol{\beta}}$, where $\hat{\boldsymbol{\beta}}$ is as given by any solution to the equations

$$\begin{bmatrix} X'V^g X & X'H_2' \\ H_2 X & 0 \end{bmatrix}\begin{bmatrix} \hat{\boldsymbol{\beta}} \\ \boldsymbol{\lambda} \end{bmatrix} = \begin{bmatrix} X'V^g \\ H_2 \end{bmatrix}\mathbf{y}, \qquad [7.1]$$

in which H_2 is a row-orthonormal matrix of latent vectors corresponding to the zero latent roots of V, and $\boldsymbol{\lambda}$ is a vector of Lagrange multipliers.

Let H^* be defined as

$$H^* = \begin{bmatrix} H_1(V^{\frac{1}{2}})^g \\ H_2 \end{bmatrix} = \begin{bmatrix} \Lambda^{-\frac{1}{2}}H_1 \\ H_2 \end{bmatrix}, \qquad [7.2]$$

where H_1 is a row-orthonormal matrix, complementary to H_2, such that $H_1VH_1' = \Lambda$, a diagonal matrix of nonzero latent roots of V, and $(V^{\frac{1}{2}})^g = H_1'\Lambda^{-\frac{1}{2}}H_1$. The notation is thus the same as that used on page 73 in describing the transformation of Rayner & Livingstone (1965), except that V is now $n \times n$.

If $\mathbf{y}$ is transformed by $\mathbf{y}^* = H^*\mathbf{y}$ to

$$\begin{bmatrix} \mathbf{y}_1^* \\ \mathbf{y}_2^* \end{bmatrix} = \begin{bmatrix} \Lambda^{-\frac{1}{2}}H_1 \\ H_2 \end{bmatrix}\mathbf{y}, \qquad [7.3]$$

then $\text{var}(\mathbf{y}_1^*) = I_r$ and $\text{var}(\mathbf{y}_2^*) = H_2VH_2' = 0$. The latter implies that, with probability one, $H_2X\boldsymbol{\beta} = H_2\mathbf{y}$, where $H_2\mathbf{y}$ is a known vector once the data have been obtained. This transformation is a special case of Rayner & Livingstone's, namely with $A = I$.

The model has thus been reduced by a nonsingular transformation to one with unit variance matrix and $n - r$ linear constraints upon $\boldsymbol{\beta}$. This is essentially the method employed by Goldman & Zelen, except that they partitioned off a zero leading submatrix of H_2X corresponding to those rows of H_2 which are orthogonal to X. Thus the $n - r$ restrictions on $\boldsymbol{\beta}$ are reduced by the number of rows of this null submatrix. Goldman & Zelen also state that the restrictions are linearly independent, which is not necessarily true.

The transformed model is

$$\left.\begin{aligned} E(\mathbf{y}_1^*) &= X^*\boldsymbol{\beta}, \quad \operatorname{var}(\mathbf{y}_1^*) = I_r \\ L\boldsymbol{\beta} &= \mathbf{c} \end{aligned}\right\} \qquad [7.4]$$

where $X^* = \Lambda^{-\frac{1}{2}}H_1X$, $L = H_2X$, and $\mathbf{c} = H_2\mathbf{y}$. Since

$$(X^*)'X^* = X'H_1'\Lambda^{-1}H_1X = X'V^gX \quad \text{(cf. Example 1.2)}$$

and

$$(X^*)'\mathbf{y}_1^* = X'H_1'\Lambda^{-1}H_1\mathbf{y} = X'V^g\mathbf{y},$$

it follows from Theorem 6.11 that the normal equations for model [7.4] are given by [7.1].

Goldman & Zelen did not derive [7.1] or give any explicit expression for $\boldsymbol{\theta}'\hat{\boldsymbol{\beta}}$, but they mentioned that it was possible to construct a system of equations in $\hat{\boldsymbol{\beta}}$ by using their methods for the linear model with given restrictions (cf. last paragraph of § 6.5). However, they did show that the BLU estimates of the set of estimable functions $\boldsymbol{\theta}'\boldsymbol{\beta}$ obtained from [7.4] by finding $\mathbf{a}_1'\mathbf{y} + \mathbf{a}_2'\mathbf{c}$ so as to minimize $\mathbf{a}_1'\mathbf{a}_1$ subject to [6.29] bear a (1, 1) relationship with the BLU estimates obtained by finding $\mathbf{w}'\mathbf{y}$ so as to minimize $\mathbf{w}'V\mathbf{w}$ subject to $\mathbf{w}'X = \boldsymbol{\theta}'$. Rohde (1964), on the other hand, reduced the model to a restricted model of the type [7.4] by means of a nonsingular transformation matrix similar to [5.14], and derived without further proof an explicit expression for the BLU estimate of $\boldsymbol{\theta}'\boldsymbol{\beta}$ by utilizing a g_3-inverse of the bordered matrix of the normal equations thereby arising (cf. the remark following Theorem 3.5). Similarly, Mitra & Rao (1968a), using the same transformation as Goldman & Zelen, state that the original problem is now reduced to one in terms of [7.4], but do not obtain explicit expressions for a BLU estimate. They rely on the proposition that under a linear model, BLU estimates of estimable functions remain invariant under a nonsingular transformation of the observations.

While we agree with Mitra & Rao, the transformation [7.3] is nevertheless somewhat special in that an original n-variate vector is transformed to an r-variate vector ($\mathbf{y}_1^*$), and in the reverse transformation the r variates are restored to the original dimension. It therefore does no harm to give a formal proof of Theorem 7.1.

Since $X'V^gX$ is positive semidefinite, it is possible to obtain a solution to [7.1] by using the expression for a g_1-inverse of a bordered matrix given in [3.43]. Thus, if

$$G_{11} = K^{g_1} - K^{g_1}X'H_2'R^{g_1}H_2XK^{g_1},$$

and

$$G_{12} = K^{g_1}X'H_2'R^{g_1},$$

where $K = X'V^g X + X'H_2'H_2 X$ and $R = H_2 X K^{g_1} X' H_2'$,

$$\boldsymbol{\theta}'\hat{\boldsymbol{\beta}} = \boldsymbol{\theta}'(G_{11}X'V^g + G_{12}H_2)\mathbf{y} = \boldsymbol{\theta}'A\mathbf{y} \qquad \text{(say)}. \qquad [7.5]$$

Now, by Theorem 6.1 $\boldsymbol{\theta}'A\mathbf{y}$ is an unbiased estimate of $\boldsymbol{\theta}'\boldsymbol{\beta}$ if and only if $XAX = X$, and this condition is independent of the variance matrix of $\mathbf{y}$. Since $G_{11}X'V^g X + G_{12}H_2 X = K^{g_1}K$ by [6.27]

$$XAX = X(G_{11}X'V^g + G_{12}H_2)X = XK^{g_1}K. \qquad [7.6]$$

Also, by [3.39] $X^*K^{g_1}K = X^*$, from which it follows that $H_1'H_1XK^{g_1}K = H_1'H_1X$. The substitution $I - H_2'H_2$ for $H_1'H_1$ now shows, since $H_2X = L$ and $LK^{g_1}K = L$ by [3.37], that $XK^{g_1}K = X$, and hence by [7.6] that A is a g_1-inverse of X, i.e. the condition for unbiasedness is met.

Further, by Theorem 6.2, $\boldsymbol{\theta}'A\mathbf{y}$ is a minimum-variance estimate of $\boldsymbol{\theta}'\boldsymbol{\beta}$ if and only if $VA'\boldsymbol{\theta}$ is a vector in the column-space of X. Now

$$\begin{aligned} VA'\boldsymbol{\theta} &= VV^gXG_{11}'\boldsymbol{\theta} + VH_2'G_{12}'\boldsymbol{\theta} \\ &= VV^gXG_{11}'X'\mathbf{a} \end{aligned} \qquad [7.7]$$

by [6.2] and since $VH_2' = 0$. But $H_2XG_{11}' = 0$ by [6.28], and therefore $H_2'H_2XG_{11}'X' = 0$, i.e. $XG_{11}'X' = XG_{11}'X'H_1'H_1$. From this it follows that $\mathrm{r}(XG_{11}'X') \leqslant \mathrm{r}(H_1'H_1) = \mathrm{r}(V)$, i.e. $XG_{11}'X' = VU$ for some matrix U; hence by [7.7] $VA'\boldsymbol{\theta} = VU\mathbf{a} = XG_{11}'X'$, and the condition for minimum variance is met.

The following corollary, which is stated without proof, relates the result of Theorem 7.1 to the customary least-squares approach for linear estimation.

Corollary 1 Equations [7.1] are obtained by minimizing the residual quadratic $(\mathbf{y} - X\hat{\boldsymbol{\beta}})'V^g(\mathbf{y} - X\hat{\boldsymbol{\beta}})$, subject to $H_2X\hat{\boldsymbol{\beta}} = H_2\mathbf{y}$.

Corollary 2 If $\mathrm{r}(X^*) = \mathrm{r}(\Lambda^{-\frac{1}{2}}H_1X) = \mathrm{r}(X)$, a solution for $\hat{\boldsymbol{\beta}}$ may be obtained by means of [3.44].

Proof: If $\mathrm{r}(X^*) = \mathrm{r}(X)$, the row-spaces of X^* and X are the same. Then H_2X is contained in the row-space of $X'V^gX$, and [7.2] may be solved by using [3.44].

Corollary 3 If X is contained in the column-space of V, the BLU estimate of $\boldsymbol{\theta}'\boldsymbol{\beta}$ is given by $\boldsymbol{\theta}'\hat{\boldsymbol{\beta}}$, where $\hat{\boldsymbol{\beta}}$ is a solution to $X'V^gX\hat{\boldsymbol{\beta}} = X'V^g\mathbf{y}$.

Proof: If $X = VT$ (say), then $H_2X = H_2VT = 0$ and the result follows.

Finally, if V is positive definite, the BLU estimate of $\boldsymbol{\theta}'\boldsymbol{\beta}$ is given by $\boldsymbol{\theta}'(X'V^{-1}X)^{g_1}X'V^{-1}\mathbf{y}$, as mentioned at the beginning of the chapter.

It is interesting to compare the results developed thus far with those presented by Khatri (1968). If F is an orthogonal complement of V, and $\tilde{V}$ is a symmetric g_1-inverse of V, Khatri shows that the BLU estimate of an estimable linear function $\boldsymbol{\theta}'\boldsymbol{\beta}$ is given by $\boldsymbol{\theta}'\hat{\boldsymbol{\beta}}$, where $\hat{\boldsymbol{\beta}}$ is as given by a solution to

$$\begin{bmatrix} X'\tilde{V}X & X'F' \\ FX & 0 \end{bmatrix} \begin{bmatrix} \hat{\boldsymbol{\beta}} \\ \boldsymbol{\lambda} \end{bmatrix} = \begin{bmatrix} X'\tilde{V} \\ F \end{bmatrix} \mathbf{y}. \qquad [7.8]$$

Khatri points out that equations [7.8] are obtained by minimizing $(\mathbf{y} - X\boldsymbol{\beta})'\tilde{V}(\mathbf{y} - X\boldsymbol{\beta})$, subject to $FX\boldsymbol{\beta} = F\mathbf{y}$.

He further mentions that any g-inverse of V may be used in the minimization without affecting the end result (cf. also Zyskind & Martin; Mitra & Rao, 1968a). The following theorem establishes that the residual quadratic does not depend on the choice of g-inverse of V.

THEOREM 7.2 If $\mathrm{E}(\mathbf{y}) = X\boldsymbol{\beta}$ and $\mathrm{var}(\mathbf{y}) = V$, $(\mathbf{y} - X\boldsymbol{\beta})'V^{g_1}(\mathbf{y} - X\boldsymbol{\beta})$ is invariant under choice of g-inverse of V.

Proof: Since $\mathrm{r}(V) = r \leqslant n$, there exists a matrix H_2, of order $(n - r) \times n$, such that $H_2 V = 0$. This implies that, with probability one, $H_2(\mathbf{y} - X\boldsymbol{\beta}) = 0$ (cf. [5.15]), and hence that $\mathbf{y} - X\boldsymbol{\beta}$ is contained in the column-space of V. Therefore, by Theorem 2.17, $(\mathbf{y} - X\boldsymbol{\beta})'V^{g_1}(\mathbf{y} - X\boldsymbol{\beta})$ is unique.

Since H_2 is an orthogonal complement of V, we may substitute $F = H_2$ in Khatri's result (it will be assumed for the rest of this section that $F = H_2$), and therefore, in minimizing $(\mathbf{y} - X\boldsymbol{\beta})'\tilde{V}(\mathbf{y} - X\boldsymbol{\beta})$ subject to $H_2X\boldsymbol{\beta} = H_2\mathbf{y}$, we are in fact minimizing $(\mathbf{y} - X\boldsymbol{\beta})'V^{g}(\mathbf{y} - X\boldsymbol{\beta})$ subject to the same restrictions, i.e. equations [7.1] and [7.8] are equivalent. However, since $X'\tilde{V}X$ is not necessarily positive semidefinite, [3.43] cannot be used to solve [7.8]. This is resolved by noting that (cf. Khatri, 1968), if $N_1 = X'V^{g}V\tilde{V}H_2'$ and $N_2 = H_2VX$, then

$$\begin{bmatrix} I & -N_1 \\ 0 & I \end{bmatrix} \begin{bmatrix} X'\tilde{V}X & X'H_2' \\ H_2X & 0 \end{bmatrix} \begin{bmatrix} I & 0 \\ -N_2 & I \end{bmatrix} = \begin{bmatrix} X'V^{g}X & X'H_2' \\ H_2X & 0 \end{bmatrix},$$

since $V = H_1'\Lambda H_1$ and $V^{g} = H_1'\Lambda^{-1}H_1$. Thus, if M and M^* denote the coefficient matrices on the left-hand sides of [7.1] and [7.8] respectively, then there exist nonsingular matrices P_1 and P_2 such that $P_1M^*P_2 = M$. It now follows that $(M^*)^{g_1} = P_2 M^{g_1}P_1$, and a g_1-inverse of M is given by [3.43].

7.3 The Zyskind–Martin method

It is apparent from the previous section that the set of equations $X'V^g X\hat{\boldsymbol{\beta}} = X'V^g\mathbf{y}$ does not, in general, lead to BLU estimates of estimable linear functions. However, Zyskind & Martin (1967) have specified a class of g_1-inverses V^* of V, such that the BLU estimate of an estimable parametric function $\boldsymbol{\theta}'\boldsymbol{\beta}$ is given by $\boldsymbol{\theta}'\mathbf{b}$, where $\mathbf{b}$ is a solution to the "general normal equations" $X'V^*X\mathbf{b} = X'V^*\mathbf{y}$. The structure of such a class of g_1-inverses is discussed below.

THEOREM 7.3 Let $\mathbf{b}$ be a solution to the equations $X'V^{g_1}X\mathbf{b} = X'V^{g_1}\mathbf{y}$. Then $\boldsymbol{\theta}'\mathbf{b}$ is an unbiased estimate of $\boldsymbol{\theta}'\boldsymbol{\beta}$ if and only if $\mathrm{r}(X'V^{g_1}X) = \mathrm{r}(X)$.

Proof:

$$\begin{aligned} \mathrm{E}(\boldsymbol{\theta}'\mathbf{b}) &= \mathrm{E}[\boldsymbol{\theta}'(X'V^{g_1}X)^{g_1}X'V^{g_1}\mathbf{y}] \\ &= \boldsymbol{\theta}'(X'V^{g_1}X)^{g_1}X'V^{g_1}X\boldsymbol{\beta}, \end{aligned}$$

and $\mathrm{E}(\boldsymbol{\theta}'\mathbf{b}) \equiv \boldsymbol{\theta}'\boldsymbol{\beta}$ if and only if $\boldsymbol{\theta}'(X'V^{g_1}X)^{g_1}X'V^{g_1}X = \boldsymbol{\theta}'$. However, for any estimable function $\boldsymbol{\theta}'\boldsymbol{\beta}$, we have $\boldsymbol{\theta}' = \boldsymbol{\alpha}'X$, whence $\boldsymbol{\theta}'\mathbf{b}$ is unbiased if and only if $\mathrm{r}(X'V^{g_1}X) = \mathrm{r}(X)$.

Corollary If $V^\ddagger$, a g_1-inverse of V, is such that $\boldsymbol{\theta}'\mathbf{b}$, where $\mathbf{b}$ is any solution to $X'V^\ddagger X\mathbf{b} = X'V^\ddagger\mathbf{y}$, is an unbiased estimate of $\boldsymbol{\theta}'\mathbf{b}$, then $\boldsymbol{\theta}'\mathbf{b}$ is unique for a given $V^\ddagger$.

Proof: As seen in the theorem, for $\boldsymbol{\theta}'\mathbf{b}$ to be an unbiased estimate of $\boldsymbol{\theta}'\boldsymbol{\beta}$, $V^\ddagger$ must be such that $\boldsymbol{\theta}'$ is in the row-space of $X'V^\ddagger X$, and by Theorem 1.6, Corollary 5, this means that $\boldsymbol{\theta}'\mathbf{b}$ is unique.

If $\boldsymbol{\theta}'(X'V^{g_1}X)^{g_1}X'V^{g_1}\mathbf{y}$ is to be a minimum-variance estimate of $\boldsymbol{\theta}'\boldsymbol{\beta}$ as well as unbiased, an additional necessary and sufficient condition is supplied by Theorem 6.2, namely that $V(V^{g_1})'X\boldsymbol{\alpha}$, where $\boldsymbol{\alpha}' = \boldsymbol{\theta}'(X'V^{g_1}X)^{g_1}$, should belong to the column-space of X.

Zyskind & Martin use these two conditions and the general form for V^{g_1} to show that the particular class of g_1-inverses exists.

Let H^* be as given in [7.2]; then by [1.11]

$$V^{g_1} \equiv (H^*)'\begin{bmatrix} I_r & U \\ Z' & Y \end{bmatrix} H^*,$$

where U, Z', and Y are arbitrary. Furthermore,

$$V(V^{g_1})' \equiv (H^*)^{-1}\begin{bmatrix} I_r & Z \\ 0 & 0 \end{bmatrix} H^*. \qquad [7.9]$$

Now, let the dimension of the intersection of the p-dimensional column-space of X and the r-dimensional column-space of V be s, and let the s independent columns of an $n \times s$ matrix Q be a basis for this intersection space. The matrix Q may be extended in any manner so that the $n \times p$ matrix $[Q \quad R]$ is a basis for the column-space of X. Zyskind & Martin show that $[Q \quad R]$ can be written as

$$[Q \quad R] = (H^*)^{-1}\begin{bmatrix} B & P_1 \\ 0 & P_2 \end{bmatrix}, \qquad [7.10]$$

where B is $r \times s$, P_1 is $r \times (p - s)$, and P_2 is $(n - r) \times (p - s)$ of rank $p - s$.

The matrix X may thus be written as $X = QC_1 + RC_2$ for some matrices C_1 and C_2, whence

$$\begin{aligned} V(V^{g_1})'X\boldsymbol{\alpha} &= V(V^{g_1})'QC_1\boldsymbol{\alpha} + V(V^{g_1})'RC_2\boldsymbol{\alpha} \\ &= QC_1\boldsymbol{\alpha} + V(V^{g_1})'RC_2\boldsymbol{\alpha}, \end{aligned}$$

since Q is contained in the column-space of V. If $V(V^{g_1})'R = 0$, then $V(V^{g_1})'X\boldsymbol{\alpha} = QC_1\boldsymbol{\alpha}$, which is a vector in the column-space of X by the construction of Q. Therefore, from [7.9] and [7.10], $V(V^{g_1})'X\boldsymbol{\alpha}$ will be contained in the column-space of X if

$$V(V^{g_1})'R = (H^*)^{-1}\begin{bmatrix} I_r & Z \\ 0 & 0 \end{bmatrix} H^*(H^*)^{-1}\begin{bmatrix} P_1 \\ P_2 \end{bmatrix} = 0,$$

i.e. if

$$P_1 + ZP_2 = 0. \qquad [7.11]$$

By Theorem 1.6, Corollary 3, the condition for equations [7.11] to admit a solution for Z is $P_1 P_2^{g_1} P_2 = P_1$, which holds since P_2 has full column-rank, and therefore $P_2^{g_1} P_2 = I$. Hence a g_1-inverse of V of the form

$$V^* = (H^*)'\begin{bmatrix} I_r & U \\ Z' & Y \end{bmatrix} H^*, \qquad [7.12]$$

where Z is any solution to [7.11] and U and Y are arbitrary, is such that $V(V^*)'X\boldsymbol{\alpha}$ is a vector in the column-space of X.

Zyskind & Martin also show that $\mathrm{r}(X'V^*X) = \mathrm{r}(X)$ if and only if $P_2'UP_1 + P_2'YP_2$ is nonsingular. However, since the objective is merely to show that V^* exists, it suffices to note that if V^* is positive definite, then $\mathrm{r}(X'V^*X) = \mathrm{r}(X)$. Since V^*, of the form

$$V^* = (H^*)'\begin{bmatrix} I_r & Z \\ Z' & (I + Z'Z) \end{bmatrix} H^*,$$

where Z is a solution to [7.11], is positive definite, it is alway possible to construct V^* to meet the rank requirement.

The following theorem has thus been proved:

THEOREM 7.4 For the linear model $\mathrm{E}(\mathbf{y}) = X\boldsymbol{\beta}$ and $\mathrm{var}(\mathbf{y}) = V$, where V is a known positive semidefinite matrix, a class V^* of g_1-inverses of V can be constructed such that, for any estimable function $\boldsymbol{\theta}'\boldsymbol{\beta}$, the BLU estimate of $\boldsymbol{\theta}'\boldsymbol{\beta}$ is given by $\boldsymbol{\theta}'\mathbf{b}$, where $\mathbf{b}$ is a solution to

$$X'V^*X\mathbf{b} = X'V^*\mathbf{y}. \qquad [7.13]$$

Zyskind & Martin purport to add that it is only for those V^* of the form [7.12], where, of course, the construction is such that $\mathrm{r}(X'V^*X) = \mathrm{r}(X)$, that equations [7.13] lead to BLU estimates. However, this cannot be correct since the condition under which V^* was derived, namely $V(V^*)'R = 0$, is only a sufficient condition for $V(V^*)'X\boldsymbol{\alpha}$ to be a vector in the column-space of X. These authors also point out that, in general, the various constructions of V^{g_1} presented in the literature do not belong to V^*.

It has already been shown in Theorem 7.3, Corollary, that $\boldsymbol{\theta}'\mathbf{b}$ is unique. In the following corollaries, all due to Zyskind & Martin, we present further properties of the "general normal equations" [7.13].

Corollary 1 For every choice of V^* the set of solutions to $X'V^*X\mathbf{b} = X'V^*\mathbf{y}$ is the same.

Proof: Let V_1^* and V_2^* be any two g_1-inverses which satisfy the requirements of Theorem 7.4. Then $\mathrm{r}(X'V_1^*X) = \mathrm{r}(X'V_2^*X) = \mathrm{r}(X)$, and there is a nonsingular matrix A such that $AX'V_1^*X = X'V_2^*X$. Let $\mathbf{b}_1$ be any solution to $X'V_1^*X\mathbf{b} = X'V_1^*\mathbf{y}$. Then

$$AX'V_1^*\mathbf{y} = AX'V_1^*X\mathbf{b}_1$$

and the right-hand side represents all BLU estimates of $AX'V_1^*X\boldsymbol{\beta}$, and hence of $X'V_2^*X\boldsymbol{\beta}$. But $X'V_2^*\mathbf{y}$ represents the set of all BLU estimates of $X'V_2^*X\boldsymbol{\beta}$, i.e. the equations $X'V_1^*X\mathbf{b} = X'V_1^*\mathbf{y}$ and $X'V_2^*X\mathbf{b} = X'V_2^*\mathbf{y}$ are equivalent.

Corollary 2 If $\mathbf{b}$ is any solution to [7.13] and H_2 is as in Theorem 7.1, $H_2X\mathbf{b} = H_2\mathbf{y}$.

Proof: By [6.2] $H_2X\boldsymbol{\beta}$ is estimable, and its BLU estimate is $H_2X\mathbf{b}$. However, $H_2\mathbf{y}$ is an unbiased estimate of $H_2X\boldsymbol{\beta}$ with zero variance matrix since $H_2 V = 0$. It therefore follows that $H_2 X\mathbf{b} = H_2 X\boldsymbol{\beta} = H_2 \mathbf{y}$.

It is evident that, for any symmetric V^* (Zyskind & Martin say positive semidefinite), the minimization of $(\mathbf{y} - X\boldsymbol{\beta})'V^*(\mathbf{y} - X\boldsymbol{\beta})$ leads to equations [7.13], and that by Corollary 2 this minimization procedure without restrictions is equivalent to the minimization of $(\mathbf{y} - X\boldsymbol{\beta})'V^*(\mathbf{y} - X\boldsymbol{\beta})$ subject to $H_2 X\boldsymbol{\beta} = H_2\mathbf{y}$. In view of Theorem 7.2 and Corollary 1 above we may now state the following corollary:

Corollary 3 The set of solutions $\mathbf{b}$ to [7.13] is identical with the set of solutions $\hat{\boldsymbol{\beta}}$ to [7.1].

This corollary therefore provides the link between the two methods of estimation discussed in the present chapter.

Zyskind & Martin have also discussed a procedure for testing hypotheses concerning solutions to equations [7.13].

Theorem 7.4 establishes the existence of normal equations with the required properties when V is singular, so that the formulation of estimation procedures in the linear model is complete. However, there remains ample scope for further research in this direction. For example, it would be desirable to have an explicit expression for V^{g_1} such that $V(V^{g_1})'X\boldsymbol{\alpha}$ is a vector in the column-space of X.

The Zyskind–Martin method is compact and is also a natural extension to the existing knowledge on estimation in linear models. It also illustrates that the lack of uniqueness of g_1-inverses (with a corresponding increase in flexibility) is in fact an advantage rather than a disadvantage. The theoretical background of the Goldman–Zelen method is, however, considerably simpler.

In recent years work has been done on conditions under which BLU estimates of estimable linear functions in the linear model with variance matrix V are equivalent to BLU estimates for linear models with alternative variance structures. In this connection the papers by Watson (1967), Zyskind (1967), and Rao (1968) may be mentioned.

CONCLUDING REMARKS

It was not to be expected that it would be possible to stop the clock while the manuscript for this monograph was being prepared. Indeed, the research of the Indian school appearing in *Sankhyā* (copies of which reach us several months after the nominal publication date) continues unabated.

C.G. Khatri has very kindly made available to us prior to publication the manuscript of a paper in which he considers probably the whole gamut of generalized inverses to date. These include a generalization of Chipman's (1964) unique g-inverse satisfying

$$AGA = A, \quad GAG = G, \quad (AG)' = V^{-1}AGV, \quad (GA)' = U^{-1}GAU,$$

where U and V are positive definite (cf. [1.2]).

In view of this sort of development, it is not easy to see where the future of g-inverses lies; certainly the notational problem is not assisted. Even the notation of Chipman (1968), who uses $A^{g_{123}}$ for our A^{g_3}, and $A^{g_{13}}$ for the "least-squares" g-inverse of Theorem 6.3, Corollary 1 (a difficulty we circumvented by means of [6.7]), is unequal to the task. Certainly it seems probable that mathematicians will devise further types of g-inverse which may or may not find application in statistics.

Generalized inverses are without doubt a powerful theoretical tool in statistical applications where singular matrices naturally occur. Although their importance is to some extent diffused in that practical solutions to computational problems were devised many years ago and continue unaltered, their contribution to the understanding of these methods is considerable. The discomfort of statisticians of an earlier generation who used the method of imposed linear restrictions (§§6.3–6.4) as a sort of *deus ex machina* should now be completely banished, and in this respect we agree fully with Searle (1967).

REFERENCES

Aitken, A.C. (1934). On least squares and linear combination of observations. *Proc. Roy. Soc. Edin.*, **55**, 12–16.

Aitken, A.C. (1937). Studies in practical mathematics. I: The evaluation, with applications, of a certain triple product matrix. *Proc. Roy. Soc. Edin.*, **57**, 172–81.

Aitken, A.C. (1945). Studies in practical mathematics. IV: On linear approximation by least squares. *Proc. Roy. Soc. Edin.*, Ser. A, **62**, 138–46.

Aitken, A.C. (1956). *Determinants and Matrices.* 9th edn, Oliver & Boyd: Edinburgh.

Albert, A. & Sittler, R.W. (1966). A method of computing least squares estimates that keep up with the data. *J. Soc. Ind. and Appl. Math.*, Ser. A, 3, 384–417.

Anderson, T.W. (1958). *Introduction to Multivariate Statistical Analysis.* Wiley: New York.

Baer, R. (1952). *Linear Algebra and Projective Geometry.* Academic Press: New York.

Ben-Israel, A. & Charnes, A. (1963). Contributions to the theory of generalized inverses. *J. Soc. Ind. and Appl. Math.*, **11**, 667–99.

Bjerhammar, A. (1958). A generalized matrix algebra. *Kunglika Teknisa Högkolans Handlingar (Trans. Roy. Inst. Technology*, Stockholm), No. 124, 1–32.

Bodewig, E. (1956). *Matrix Calculus.* North Holland: Amsterdam.

Boot, J.C.G. (1963). The computation of the generalized inverse of singular or rectangular matrices. *Amer. Math. Monthly*, **70**, 302–3.

Bose, R.C. (1944). The fundamental theorem of linear estimation (abstract). *Proc. 31st Indian Science Cong.*, **4**, 2–3.

Bose, R.C. (1959). Unpublished lecture notes on analysis of variance. Univ. North Carolina, Chapel Hill.

Chipman, J.S. (1964). On least squares with insufficient observations. *J. Amer. Statist. Assoc.*, **59**, 1078–1111.

Chipman, J.S. (1968). Specification problems in regression analysis. *Proc. Symposium on Theory and Application of Generalized Inverses of Matrices.* Mathematics Series No. 4, Texas Technological College, Lubbock, Texas.

Chipman, J.S. & Rao, M.M. (1964). Projections, generalized inverses and quadratic forms. *J. Math. Anal. and Applications*, **9**, 1–11.

Cline, R.E. (1964a). Representations for the generalized inverse of a partitioned matrix. *J. Soc. Ind. and Appl. Math.*, **12**, 588–600.

Cline, R.E. (1964b). Note on the generalized inverse of the product of matrices. *SIAM Rev.*, **6**, 57–8.

Desoer, C.A. & Whalen, B.A. (1963). A note on pseudoinverses. *J. Soc. Ind. and Appl. Math.*, **11**, 442–7.

Dwyer, P.S. (1958). Generalizations of a Gaussian theorem. *Ann. Math. Statist.*, **29**, 106–17.

Englefield, N.J. (1966). The commuting inverses of a square matrix. *Proc. Camb. Phil. Soc.*, **62**, 667–71.

Erdelyi, I. (1966). On the "reverse order law" related to the generalized inverse of matrix products. *J. Assoc. Computing Mach.*, **13**, 439–43.

Fox, L. (1950). Practical methods for the solution of linear equations and the inversion of matrices. *J. Roy. Statist. Soc.*, Ser. B, **12**, 120–36.

Fox, L. (1964). *An Introduction to Numerical Linear Algebra.* Clarendon Press: Oxford.

Fox, L. & Hayes, J.G. (1951). More practical methods for the inversion of matrices. *J. Roy. Statist. Soc.*, Ser. B, **13**, 83–91.

Goldman, A.J. & Zelen, M. (1964). Weak generalized inverses and minimum variance linear unbiased estimation. *J. Res. Nat. Bur. Stand.*, B, Mathematics and Mathematical Physics, **68B**, 151–72.

Golub, G.H. (1969). *Matrix decompositions and statistical calculations.* Technical Report No. CS 124, Computer Science Department, Stanford Univ.

Golub, G.H. & Kahan, W. (1964). *Calculating the singular values and pseudo-inverse of a matrix.* Technical Report No. CS 8, Computer Science Division, Stanford Univ.

Graybill, F.A. (1961). *An Introduction to Linear Statistical Models*, **1**. McGraw-Hill: New York.

Graybill, F.A., Meyer, C.D. & Painter, R.J. (1966). Note on the computation of the generalized inverse of a matrix. *SIAM Rev.*, **8**, 522–4.

Greville, T.N.E. (1959). The pseudoinverse of a rectangular matrix and its application to the solution of systems of linear equations. *SIAM Rev.*, **1**, 38–43.

Greville, T.N.E. (1960). Some applications of the pseudoinverse of a matrix. *SIAM Rev.*, **2**, 15–22.

Greville, T.N.E. (1961). Note on fitting functions of several independent variables. *J. Soc. Ind. and Appl. Math.*, **9**, 109–15.

Greville, T.N.E. (1962). Further remarks on the pseudoinverse of a matrix. Unpublished address delivered at the University of Michigan.

Greville, T.N.E. (1966). Note on the generalized inverse of a matrix product. *SIAM Rev.*, **8**, 518–21.

Harris, W.A. & Helvig, T.N. (1966). Marginal and conditional distributions for singular distributions. *Publ. Res. Inst. Math. Scientists*, Kyoto Univ., Ser. A, **1**, 200–4.

Healy, M.J.R. (1968a). Multiple regression with a singular matrix. *Applied Statist.*, **17**, 110–17.

Healy, M.J.R. (1968b). Triangular decomposition of a symmetric matrix (Algorithm AS 6). *Applied Statist.*, **17**, 195–7.

Healy, M.J.R. (1968c). Inversion of a positive semidefinite symmetric matrix (Algorithm AS 7). *Applied Statist.*, **17**, 198–9.

Hsu, P.L. (1946). On a factorization of pseudo-orthogonal matrices. *Q. J. Math.*, Oxford Ser., **17**, 162–5.

John, P.W.M. (1964). Pseudo-inverses in the analysis of variance. *Ann. Math. Statist.*, **35**, 895–6.

Kempthorne, O. (1952). *Design and Analysis of Experiments*. Wiley: New York.

Khatri, C.G. (1962). Conditions for Wishartness and independence of second degree polynomials in a normal vector. *Ann. Math. Statist.*, **33**, 1002–7.

Khatri, C.G. (1963). Further contributions to Wishartness and independence of second degree polynomials in normal vectors. *J. Indian Statist. Assoc.*, **1**, 61–7.

Khatri, C.G. (1964). Distribution of the "generalized" multiple correlation matrix. *Ann. Math. Statist.*, **35**, 1801–6.

Khatri, C.G. (1968). Some results for the singular normal multivariate regression models. *Sankhyā*, Ser. A, **30**, 267–80.

Laha, R.G. (1956). On the stochastic independence of two second-degree polynomial statistics in normally distributed variates. *Ann. Math. Statist.*, **27**, 790–6.

Lanczos, C. (1958). Linear systems in self-adjoint form. *Amer. Math. Monthly*, **65**, 665–79.

Lewis, T.O. & Odell, P.L. (1966). A generalization of the Gauss–Markov theorem. *J. Amer. Statist. Assoc.*, **61**, 1063–6.

Lucas, H.L. (1962). Unpublished lecture notes on the linear model and its analysis. Univ. North Carolina, Raleigh.

Marsaglia, G. (1964). On conditional means and covariances of normal variables with singular variance matrix. *J. Amer. Statist. Assoc.*, **59**, 1203–4.

Mirsky, L. (1955). *An Introduction to Linear Algebra*. Clarendon Press: Oxford.

Mitra, S.K. (1968a). On a generalized inverse of a matrix and applications. *Sankhyā*, Ser. A, **30**, 107–14.

Mitra, S.K. (1968b). A new class of g-inverse of square matrices. *Sankhyā*, Ser. A, **30**, 323–30.

Mitra, S.K. & Rao, C.R. (1968a). Some results in estimation and tests of linear hypotheses under the Gauss–Markoff model. *Sankhyā*, Ser. A, **30**, 281–90.

Mitra, S.K. & Rao, C.R. (1968b). Simultaneous reduction of a pair of quadratic forms. *Sankhyā*, Ser. A, **30**, 313–22.

Moore, E.H. (1920). On the reciprocal of the general algebraic matrix. (Abstract.) *Bull. Amer. Math. Soc.*, **26**, 394–5.

Moore, E.H. (1935). General Analysis. *Mem. Amer. Phil. Soc.*, **1**, 197.

Osborne, E.E. (1965). Smallest least square solutions of linear equations. *J. Soc. Ind. and Appl. Math.*, Ser. B, **2**, 300–7.

Pearl, M.H. (1966). On generalized inverses of matrices. *Proc. Camb. Phil. Soc.*, **62**, 673–7.

Penrose, R. (1955). A generalized inverse for matrices. *Proc. Camb. Phil. Soc.*, **51**, 406–13.

Penrose, R. (1956). On best approximate solutions of linear matrix equations. *Proc. Camb. Phil. Soc.*, **52**, 17–19.

Plackett, R.L. (1950). Some theorems in least squares. *Biometrika*, **37**, 149–57.

Plackett, R.L. (1960). *Principles of Regression Analysis*. Clarendon Press: Oxford.

Price, C.M. (1964). The matrix pseudoinverse and minimal variance estimates. *SIAM Rev.*, **6**, 115–20.

Quenouille, M.H. (1950). Computational devices in the application of least squares. *J. Roy. Statist. Soc.*, Ser. B, **12**, 256–72.

Rado, R. (1956). Note on generalized inverses of matrices. *Proc. Camb. Phil. Soc.*, **52**, 600–1.

Rao, C.R. (1946). On the linear combination of observations and the general theory of least squares. *Sankhyā*, **7**, 237–56.

Rao, C.R. (1955). Analysis of dispersion for multiply classified data with unequal numbers in cells. *Sankhyā*, **15**, 253–80.

Rao, C.R. (1962). A note on a generalized inverse of a matrix with application to problems in mathematical statistics. *J. Roy. Statist. Soc.*, Ser. B, **24**, 152–8.

Rao, C.R. (1965). *Linear Statistical Inference and its Applications*. Wiley: New York.

Rao, C.R. (1966). Generalized inverse for matrices and its applications in mathematical statistics. Contribution to *Research Papers in Statistics* (Festschrift for J. Neyman). Wiley: New York.

Rao, C.R. (1967). Calculus of generalized inverses of matrices. Part I: General theory. *Sankhyā*, Ser. A, **29**, 317–42.

Rao, C.R. (1968). A note on a previous lemma in least squares and some further results. *Sankhyā*, Ser. A, **30**, 259–66.

Rayner, A.A. & Livingstone, D. (1965). On the distribution of quadratic forms in singular normal variates. *South African J. Agric. Sci.*, **8**, 357–69.

Rayner, A.A. & Pringle, R.M. (1967). A note on generalized inverses in the linear hypothesis not of full rank. *Ann. Math. Statist.*, **38**, 271–3.

Rohde, C.A. (1964). Contributions to the theory, computation and application of generalized inverses. Mimeo No. 392, Institute of Statistics, Univ. North Carolina, Raleigh.

Rohde, C.A. & Harvey, J.R. (1965). Unified least squares analysis. *J. Amer. Statist. Assoc.*, **60**, 523–7.

Roy, S.N. (1953). Some notes on least squares and analysis of variance. Mimeo No. 81, Institute of Statistics, Univ. North Carolina.

Rushton, S. (1951). On least squares fitting by orthonormal polynomials using the Choleski method. *J. Roy. Statist. Soc.*, Ser. B, **13**, 92–9.

Rust, B., Burrus, W.R. & Schneeberger, C. (1966). A simple algorithm for computing the generalized inverse of a matrix. *Comm. Assoc. Computing Mach.*, **9**, 381–5, 387.

Scheffé, H. (1959). *The Analysis of Variance*. Wiley: New York.

Scroggs, J.E. & Odell, P.L. (1966). An alternative definition to the pseudo-inverse of a matrix. *J. Soc. Ind. and Appl. Math.*, **14**, 796–810.

Searle, S.R. (1965). Additional results concerning estimable functions and generalized inverse matrices. *J. Roy. Statist. Soc.*, Ser. B, **27**, 480–90.

Searle, S.R. (1966). *Matrix Algebra for the Biological Sciences (including Applications in Statistics)*. Wiley: New York.

Searle, S.R. (1967). The use of generalized inverse matrices in teaching statistics. Paper delivered at the VI International Biometric Conference, Sydney, Australia.

Shanbhag, D.N. (1968). Some remarks concerning Khatri's result on quadratic forms. *Biometrika*, **55**, 593–5.

Sheffield, R.D. (1958). A general theory for linear systems. *Amer. Math. Monthly*, **65**, 109–11.

Tewarson, R.P. (1967–8). A computational method for handling generalized inverses. *Computer J.*, **10**, 411–13.

Urquhart, N.S. (1969). The nature of the lack of uniqueness of generalized inverse matrices. *SIAM Rev.*, **11**, 268–71.

Watson, G.S. (1967). Linear least squares regression. *Ann. Math. Statist.*, **38**, 1679–99.

Wilkinson, G.N. (1958). Estimation of missing values for the analysis of incomplete data. *Biometrics*, **14**, 257–86.

Yates, F. (1933). The principles of orthogonality and confounding in replicated experiments. *J. Agric. Sci.*, **23**, 108–45.

Yates, F. (1934). The analysis of multiple classifications with unequal numbers in the different classes. *J. Amer. Statist. Assoc.*, **29**, 51–66.

Yates, F. & Hale, R.W. (1939). The analysis of Latin squares when two or more rows, columns, or treatments are missing. *J. Roy. Statist. Soc.*, Suppl., **6**, 67–79.

Zelen, M. & Federer, W.T. (1965). Application of the calculus for factorial arrangements. III: Analysis of factorials with unequal numbers of observations. *Sankhyā*, Ser. A, **27**, 383–400.

Zyskind, G. (1964). Topics in general linear models theory. Contribution to *Research on Analysis of Variance and Related Topics*, Aerospace Research Laboratories, Technical Report 64–193, Wright–Patterson Air Force Base, Ohio.

Zyskind, G. (1967). On canonical forms, non-negative covariance matrices and best and simple least squares linear estimators in linear models. *Ann. Math. Statist.*, **38**, 1092–1109.

Zyskind, G. & Martin, F.B. (1967). A general Gauss–Markoff theorem for linear models with arbitrary non-negative covariance structure. Unpublished MS.

INDEX